John Charles Bucknill

Habitual Drunkenness and Insane Drunkards

John Charles Bucknill

Habitual Drunkenness and Insane Drunkards

ISBN/EAN: 9783337371302

Printed in Europe, USA, Canada, Australia, Japan

Cover: Foto ©berggeist007 / pixelio.de

More available books at **www.hansebooks.com**

HABITUAL DRUNKENNESS

AND

INSANE DRUNKARDS.

BY

JOHN CHARLES BUCKNILL, M.D., Lond., F.R.S.,

FELLOW OF THE ROYAL COLLEGE OF PHYSICIANS;
LATE LORD CHANCELLOR'S VISITOR OF LUNATICS.

London:
MACMILLAN AND CO.
1878.

PREFACE.

ON the second reading of the Habitual Drunkards' Bill in the House of Commons, July 3rd, 1878, the introducer of the measure, Dr. Cameron, member for Glasgow, referred to the following articles in these terms :—

" The articles which had been written by Dr. Bucknill, the Lord Chancellor's Visitor, as the result of his visit to the United States, were marked by strong prejudice, as when he went so far as to say that he was ' out of all patience with maudlin sentimentality. In addressing the House he had endeavoured to abstain from that. (Cheers.) Dr. Bucknill suggested that the best thing that could happen was that the drunkard should ruin himself and his property pass into worthier hands. Dr. Bucknill's statements called forth the most conclusive replies. An institution that he reported closed through failure was closed simply because the State aid that had been expected was not forthcoming. It was said that Dr. Bucknill's reports were founded on superficial and limited information, and that he had not visited asylums the success of which had surpassed the wildest dreams of enthusiastic supporters. Careful inquiry showed that 60 per cent. of those who had passed through them continued to be sober persons."

To this attack the author replied in the *Times* by the letter which will be found at page 98, and to Dr. Cameron's reply in the same journal by the letter which will be found at page 101, and, so far as the author is personally concerned, he deems those letters an ample justification. But, on reconsidering these articles thus impugned, the author is led to believe not only that they bear internal evidence that they were written without prejudice, but that they afford an unbiassed consideration of the social and medical bearings of drunkenness, which is not undeserving of being placed before the public in a collected form. Perhaps his opinions would have been better appreciated had they been recast into one connected paper; but, under the circumstances, he has preferred to republish them, with all their faults, in the form and in the order in which they were written.

The author has not willingly entered upon this controversy: he has been drawn or driven into it. Interested as a mental physician in drunkenness as one of the great factors of insanity, in his visit to the United States in 1875, he made inquiries on the subject for the sole purpose of informing and satisfying his own mind on a most important matter closely connected with his business in life. His inquiries were made for the sole purpose of eliciting truth, and he ventures to affirm that none of the gentlemen of whom they were made will aver that they were either superficial or negligent. Others there were whose testimony he had been taught by their own countrymen to discredit, and of these he did not seek information. They have had other opportunities of giving it to the world, of which they have amply availed themselves; but neither the author nor the public have any means of testing the accuracy of their statements. When lunatics are discharged from an asylum as cured, there is some probability that they are cured, at least for a time, as there is a certain ascertainable difference between a man who has been insane and is cured, and one who has been insane and is not cured. But, notwithstanding the information given to the House of Commons, that, with regard to a drunkard, "it could be known by the hands that all danger was removed, and a cure had been effected," it must be admitted that, with regard to propensities, men do not wear the heart upon the sleeve, or even on the hands, and it will be doubted whether the discrimination of the most acute observers can tell which of two sober men is likely to drink too much whiskey when he can get it. Not only did the author visit all the inebriate asylums and homes he could hear of in the eastern States of America, with the exception of one small private institution, but he made diligent inquiries of the numerous able and experienced physicians, by whom he was received with frank friendliness in that country; and he also listened to and participated in a full and able discussion of the whole subject by the Association of Medical Officers of Asylums for the Insane, who in America, as in this country, are brought closely into relation with habitual drunkenness as a cause of insanity. His remarks on this occasion, reprinted from the Proceedings of the Association, will form a fitting introduction to this series of papers, and will prove to any candid mind how far removed he was from making his inquiries in any spirit of prejudice. There is indeed an underlined passage in these remarks which might even justify Dr. Cameron and his friends in claiming the author as a coadjutor in their cause. Nor would he refuse were the truthfulness of their assertions of cure substantiated, and their scheme of action not debased by the greed of private gain. But in this very debate, in which a great number of the most experienced mental physicians in America took part, no tittle of evidence was forthcoming that any considerable proportion of habitual drunkards were cured, while, on the contrary, Dr. John Gray, the Superintendent of the Asylum for the State of New York, asserted :—" I can recall some very remarkable cases of restoration from that habit, lasting eight, ten, or twenty years, that is, from the time of their discharge from the asylum to the present ; but *I can count them all upon my fingers.* The great majority of those who came through their own will, and seemed to have been strong when discharged, have in the main returned to drinking again." Dr. Conrad, the Superintendent of the Maryland Asylum, said :—" We have one hall devoted to inebriates or dipsomaniacs. The experience I have had in the hospital has been confined to a class known as dipsomaniacs. Many have been confined to the hospital for several years, scarcely making an endeavour to withhold from drinking for three days. .... The subject of their treatment is a matter in which I am interested, but I do not know of a single case where a cure has been effected by confinement."

Dr. Landor, the Superintendent of the Asylum for the Insane for the Province of Ontario, Canada, said :—" The Government of Ontario, more than two years ago, began to build an asylum for inebriates, which will be finished in the course of a few months, and will hold a hundred people. The Government of Ontario doubts the efficacy of admitting patients into an asylum in this way, *on account of the failure in the States*. They have hesitated from what has been done in the States, by the discharge of men as cured, while many of those discharged have returned shortly, like the sow to the mire. Therefore our Government contemplates making this new institution into a hospital for lunatics instead of drunkards."

Dr. Walker, the Superintendent of the Boston Asylum and the Vice-President of the Association, said : " They [the Legislatures] have erected inebriate asylums and they have failed to cure the cases placed therein. .... We have two such institutions in Boston, conducted by reformed inebriates, both of them honest, intelligent, and enthusiastic men. I asked one of them, 'What is necessary to reform these patients?' He answered me, 'Practical operative religion.' Without that they could do nothing whatever. In a few days after this, I went to one with Dr. Bucknill. They were opening the meeting by the singing of temperance songs and had religious exercises. Without this religion there is no hope for them. We know that both confinement and labour have failed; both are good together, but alone without religion they can do no good whatever."

Dr. Kirkbride, the superintendent of the Pennsylvania Hospital for the Insane, the venerated chief of mental physicians in the United States, said: "From what has been said here, and what is said elsewhere, joined to many cases which have come under my own observation, I do not think it will be too much to say that these inebriate asylums have been failures. They certainly have not done what was expected of them when they were established."

The superintendents of asylums for the insane in America have large opportunities of forming an opinion on the matters in question because, as the President of the Association stated, medical men all over the country freely give certificates of insanity, upon which inebriates are sent in considerable numbers to asylums for the insane.

The author heard opinions fully as strong expressed by private persons, and he cannot understand what, except ignorance, could have induced Dr. Cameron to have impugned his own moderate and unbiassed statements, which did not accuse these asylums of failure, but only of such a system of management observed by himself as appeared to give but little promise of success.

Ward's Island, indeed, has been given up to reprobation, and the reader will perhaps suspect from the words of Dr. Walker that there was more singing than work at Boston, especially as the religious and reformed inebriate who conducted this institution confessed that he had no means of preventing his inmates from getting whiskey, and that there was a spirit store next door. Binghampton, the great whiskey cure of New York State, has been a scandal from its inception, when an English philanthropist, whose name also was Walker, had collected a vast sum from the benevolent, which he had expended in a magnificent site and building, which he said was his own when the philanthropists wished to oust him, and Mr. Walker had to be paid for it a second time. Then came Dr. Dodge and Dr. Day, the other reformed inebriate whom Dr. Walker, of Boston, referred to, then Dr. Congdon, whom I examined, followed by Dr. Crother, who writes to Mr. Cameron that 60 per cent. of the inmates were cured in an institution where there was neither moral nor medical treatment, and where it was freely admitted that the inmates could get as much whiskey as they liked by walking to the outskirts of the town just outside their own grounds. So much inaccuracy of statement has been made in this discussion by Dr. Cameron and his friends that we cannot at once accept the report of Dr. Crother as far as it goes, namely, that—

"Last year a careful examination was instituted into the history of patients who had been here for treatment over five years ago. Letters were addressed to their friends inquiring into their present condition, and the answers revealed the startling fact that over 60 per cent. of those who were under treatment here for four or five months, were sober, and continued reformed men."—*Times*, July 11th.

But even if the friends of the Binghampton patients did so answer Dr. Crother's inquiry, it would not prove that their account was correct, and nothing could be accepted as a satisfactory verification of the astounding announcement that 60 per cent. of the inmates in this rotten institution were permanently reformed, except an independent inquiry into their subsequent history. Since Dr. Crother's communication, he also has been superseded at Binghampton, and by a gentleman who bears a reputation for honesty and ability. Would Dr. Kitchen undertake the trouble and the odium of making a real investigation into the statistics of this place which to many minds hitherto appears to rear its head with the moral attributes of London's tall column?

After having inquired into the treatment of drunkards in America for the simple purpose of information, the

Author owes it to himself, after the virulent attacks that have been made upon him, to describe how dispassionately and unintentionally he was drawn into controversy on the subject. He had not entertained the slightest intention of publishing his observations and opinions, but some time after his return from America, on attending at the annual temperance meeting at Rugby, he was induced to make some observations on the subject, the report of which came under the notice of Dr. Clouston, one of the editors of the *Journal of Mental Science*, and this gentleman was so struck with them that he addressed to the Author a letter containing a very fair and able criticism, to which the Author responded in the second article of this collection, which will be found to contain his opinions and the reasons on which he holds them stated frankly, but he trusts also clearly and calmly. He much wishes that the discussion, if it were to proceed further, could have been kept in this even tenor of argument, but unfortunately a slight reference had been made to a gentleman who was a red-hot partizan of the drunkard's legislation, who rushed in with an angry and illogical attack upon the Author, thus drawing upon himself the sharp rebuke of the third article, which the Author would willingly leave in the leaves of a specialist journal, had he not resolved in reply to Dr. Cameron's accusation, to republish all and the whole of what he has written on the subject.

The Author has no excuse to make for a passage in the fourth article, which has been repeatedly quoted against him, if it be not true. But if it be true, there can surely be no blame in stating forcibly a conclusion of social science, which is, that in these later ages the uncontrolled indulgence of men's propensities tends to eliminate the unfittest, and thus to preserve or improve the best qualities of the race. In the ages of personal conflict, it was the weak or the slow man who went down early in the struggle for existence; in these times of peaceful conflict it is the foolish or the vicious who succumb; and, now there seems to be no element of survivalism more potent than temperance. Let those who cannot disprove it forbear to sneer at the exposition of a most wholesome truth.

In the fifth article the Author has been fortunate to escape from the atmosphere of controversy into which he had been dragged. Written hurriedly, on a sudden demand for a topic of scientific discusssion, he cannot but think that it is of value not only as an attempt further to define the relations of drink and insanity, but also as an early statement of the possible action of alcohol in averting disease. The views taken in this paper with regard to the atriptic action of alcohol in averting mental disease have recently been followed up by Dr. W. Farr in a letter to the Registrar-General on the possible action of alcohol in averting zymotic disease. For his article the Author was honoured by the *Temperance Review* with the title of the Apostle of Moderate Drinking which he is fain to set against Dr. Cameron's strictures on his harshness to drunkards, as another instance that zealots detest a man of moderate views far more than they do an opposite zealot. To the first they are antipathetic, with the other they are in the full swing of sympathy, however hostile they may be on the less important matter of opinion. If Dr. Cameron is right and drunkenness is a disease, the temperance movement is a cruel persecution of the innocent. If Sir Wilfred Lawson is right, the indulgent treatment of drunkards is an absurd and mischievous offence against society. The connivance of temperance men at the disease-heresy of drunkenness, is, however, confined to this country, for in America it has been earnestly and honestly repudiated.

The American Association for the Cure of Inebriates, propounds the broad principle that drunkenness is disease, and has, in support of this dogma, gone the length of rejecting the report of the best inebriate home in the whole country, because it was established for the reform, and not the cure of drunkenness. Dr. Cameron, however, has improved even upon the absurd idea of this general view, by asserting to the House of Commons that drunkenness is not only a disease, but that it is what is called an organic disease, in which there are "certain structural alterations, especially in the nervous system." "The result of these deteriorations of structure rendered the patient unable to resist that craving" for stimulants, &c.[1] And he

[1] "The evidence showed that there existed a very close analogy between habitual drunkenness and the various forms especially of mental disease.

proceeds to affirm that these deteriorations of structure have been permanently removed in a certain hospital too remote for the statement to be verified, at the rate of 60 per cent. of the patients discharged.

Seeing the manner in which the author has spoken of inebriate asylums in his own country, he has been somewhat surprised at the acrimony with which he has been attacked for the statements he has made respecting the American institutions. He has visited six of these institutions in England and Scotland, and he has had every reason to believe that with much less pretence, much more earnest and good work is being done in them than in the American establishments. The only explanation which occurs to him is that these attacks come from members of the "Society for Promoting Legislation for the Control and Care of Habitual Drunkards," some of whose members are interested in procuring legal powers of detention which they have erroneously supposed to exist in America.

The English asylums for drunkards are private speculations, and, in default of powers of detention, they are said to be unremunerative. Give us—say members of this philanthropic society—power of

---

Prolonged indulgence in alcohol or other stimulants after a time—longer in some cases, shorter in others—brought about certain structural alterations, especially in the nervous system. The result of these deteriorations of structure were, first, a depression of vital force, giving rise to a craving for stimulants; and second, a depression of the force of intellect and the will, which rendered the patient unable to resist that craving."—From Dr. CAMERON's Speech; see the *Times*, July 4th, 1878.

arrest and compulsory detention, and we shall keep our asylums full, and make them as profitable as private lunatic asylums. The Author has too plainly pointed out that in this agitation the angel wings of philanthropy are not long enough to hide the cloven foot of self-interest. Hence all these corrosive tears which have been shed upon him. That any society of which the Earl of Shaftesbury is president can make any permanent deviation from the paths of the highest philanthropy the Author does not for one moment believe, but that pecuniary advantage to individuals has been a prime motor in the minds of many leaders of this agitation cannot be doubted, seeing that some of them have had the courage to avow it publicly and in print.[1]

That Dr. Cameron, entertaining the views he does about the curability of structural alterations of the nervous system, should differ from the Author, is not surprising; but that he should single him out for attack in the House of Commons may perhaps also, to some extent, be accounted for by the manner in which the Author exposed the manifold follies and tyrannies of last year's Bill, when invited to meet Dr. Cameron for its discussion at the Metropolitan Branch of the British Medical Association, at which discussion he had the audacity to suggest that the Member for Glasgow might usefully employ his influence in promoting the execution of the existing law among

---

[1] See Appendix to Bluebook on Habitual Drunkards, and report of discussions at Edinburgh meeting of British Medical Association.

his constituents, to whom the Author described a visit of investigation in 1876 in company with the Sheriff of Lanarkshire, in which, on Saturday night and Sunday morning, he saw, with amazement and horror, a drunken population: the drink-shops and lodging-houses teeming, and the streets crowded with drunken men and women, and even children; the numerous police-cells occupied by comatose drunkards, and the police surgeon attending, through the long hours of the night, to drunken accidents; and—strange necessity!—the police constables taking no heed of simple drunkenness, and admitting, to the sheriff and his companion, that they never did interfere so long as the drunkard was peaceable, and could keep on his legs. When he fell helpless on the pavement they put him in a wheelbarrow stretcher and trundled him off to the cells. Dr. Cameron did not relish this description, and endeavoured to deny it; but it may still be repeated by any one who will spend the last and the first hours of the week in the wynds of Glasgow. And Dr. Cameron was reminded that another observer had described the same orgie in almost identical terms in the *Daily News*, and that he had been insulted and threatened, and had been compelled to leave Glasgow, on account of his truthful audacity. As it is from the occasional drunkard that the habitual drunkard is manufactured, why does not Dr. Cameron attack habitual drunkenness at its source? even at the risk of offending the bulk of his constituents.

That remainder biscuit which is left to Dr. Cameron after his legislative voyage it would, perhaps, be cruel to cast away; yet it is full of weevil, which may, perchance, contaminate sounder food. It is scarcely denied that, if there were no vested interests to consider, private lunatic asylums could not be permitted; and yet this fragment of legislation would establish a new class of asylums, in which the liberty of the subject would be submitted to the determinations, biased by self-interest, of private adventurers, using that term in the sense in which it is employed to distinguish adventure schools from public schools. As the *Spectator* has pointed out, if such a private adventurer kept an alleged drunkard six or twelve months in his asylum when he ought to have been discharged in three, a grave injustice would be committed. And it must not be overlooked that the inebriate doctor who keeps an adventure asylum would have much greater opportunity of committing such an act of injustice than the mental doctor who keeps an adventure lunatic asylum; for, while there is a marked distinction between the man who remains insane and the one who has recovered, there are no means of distinguishing the man who is still inclined to get drunk, from the man who has firmly and successfully determined not to do so.

Under this Bill an attesting justice or commissioner to administer oaths is to state in writing that the applicant understood the effect of his application, &c., that is, the surrender of his liberty; but this a man might do while still incapable from the stupidity of a debauch. It ought at least to be attested upon medical examination that the man was sane and sober at the time when he is persuaded to take so important a step as to turn the key of a prison upon himself for a twelvemonth.

There cannot be a doubt that, unless stringent measures are taken to prevent it, these new asylums will be resorted to as places of detention for real lunatics; and the Author is able to assert, from his own personal observation, that they are already resorted to for this purpose. Last year he examined one of these places by order of the Lord Chancellor, and, while entertaining doubts about the sanity of other inmates, he found one lady whose state of lunacy was obvious and unquestionable, and for whose detention without certificates the keeper of the asylum was prosecuted and punished.

The spirit of Dr. Cameron's legislative proposals is marked by the monstrous penalties which he proposes to attach to any offence against this statute.

If a lunatic escapes from an asylum there is no penalty on any one who harbours him, and if he evades recapture for seven days his certificates must be renewed. If even a felon escapes from prison, it is no offence to harbour him; but if any person, the man's wife, for instance, should conceal or harbour an habitual drunkard who has escaped from his private gaol, that person will be liable to a penalty of twenty pounds, or to imprisonment, with hard labour, for six months. Could not Dr. Cameron bring a little of the repressive temper under which such penalties were conceived to bear upon the rampant vice of the town he represents? But the Bill can do little harm, because it is so small a measure; at all events give this inch of legislation a trial. This is said to one side; and to the other, it is urged, only let us get in the thin end of the wedge, and we shall eventually obtain all we desire. It is a good example of crocheteering legislation. An agitation; an association; a philanthropic balloon, with cordage of self-interest; a grand promise, and an organised pressure upon members who keep their heads, but not their patience, and at last give way, like a certain judge, saying, "It is a small thing; let them try."

The great mistake which Dr. Cameron and his friends make is in refusing to recognise the obvious and undoubted fact that there are two distinct kinds of drunkards—the habitual and the insane drunkards—who must be dealt with by different methods; the one form of drunkenness being a mere vice which may be reformed by moral methods, and of which we have seen in our own days a vast and sudden reformation under the influence of a devoted and eloquent priest; and a more slow, but also more trustworthy, reformation still proceeding by various ways, of which the highroad of the temperance associations attracts the most notice. This kind of drunkenness is too widespread to be dealt with in asylums, or brick and mortar institutions of any kind. Neither can it be cured by any form of treatment, for it is not a disease. The other kind of habitual drunkenness is a morbid condition, and is, in fact, a form of insanity, and no one has described it better than Dr. Cameron himself,[1] his only grave error being the assertion that a large percentage of such cases can be cured. The Author is convinced that he speaks with the concurrence of all well-informed members of his profession when he asserts that a man who has be-

[1] "He would not discuss the question whether dipsomania was a vice or a disease; but there could be no doubt that habitual drunkenness closely approximated to insanity. The hereditary tendency to lunacy was also frequently the effect of excessive drunkenness. In some cases men who had hitherto been sober and steady became suddenly intemperate after a sunstroke or after receiving a blow or wound. There was a monomaniacal character about the offences committed by many drunkards in their cups. From 1844 to 1865 one woman was committed 137 times for being drunk, in which condition she invariably smashed windows, and at last she drowned herself. An old soldier who was wounded in the head stole nothing but Bibles, and was transported for the seventh theft. Another man stole spades, a woman shoes, and a man was transported for the seventh theft of a tub, although tubs had nothing to do with his occupation or prospects. In many cases lunacy was attributed to intemperance, and men were committed as lunatics. Deprived of drink, they speedily recovered and were liberated. Taking to drink again, they had to be locked up; and these alternations recurred until they were no longer deemed sane, when their aberrations ceased and they were subjected to prolonged detention. This was done illegally now, and the fact showed in what direction cure was looked for. Still more did the proposal of the Scotch Commissioners in Lunacy, when the Criminal Lunatics Amendment Bill was before Parliament—that in cases where it was shown that lunacy had resulted from drinking, there should be power to detain patients for twelve months after their recovery. The proposal proved the belief of the Commissioners that habitual drunkenness depended to a large extent upon a diseased condition of the system, and that it required long and special treatment to enable the system to recover its tone and subdue the morbid craving for drink. When cases were treated on this theory, a large percentage of cures was effected."—*The Times*, July 4th.

come a drunkard after a sunstroke or a blow on the head, his craving for drink being accompanied by such other indications of mental infirmity as Dr. Cameron has specified, and in whom Dr. Cameron is perfectly right in supposing that there are structural alterations of the nervous system, must, under any methods of treatment at present known to us, be looked upon as a lunatic, presenting an exceedingly small probability of permanent cure. But, curable or not, such a person cannot, and ought not to be dealt with by moral methods alone. He is a diseased person, needing medical care and treatment, for which purpose he must be subject to compulsory arrest and detention. That peculiarity of this malady which permits long remissions of the more marked symptoms renders the insane drunkard an inconvenient inmate of ordinary lunatic asylums, and therefore the proper method of dealing with such cases would be to establish public Hospitals for Insane Drunkards, with salaried officers under the control of Boards of Governors and the authority of the Commissioners in Lunacy. Much stress has been laid upon the value of classification within asylums; that is to say, the classification of the different kinds of lunatics within the same asylum; but a still more efficient and valuable form of classification would be to provide asylums of different character for lunatics of different kinds. The so-called Criminal Lunatic Asylum, for lunatics who by law are not criminals, and the Asylum for Insane Convicts who are criminals, are examples in active operation, and Hospitals for Insane Drunkards would, the Author believes, meet the reasonable wishes of all disinterested supporters of Dr. Cameron's legislative proposals. It is not clear that any extension of the existing Lunacy Laws would be needed for the proper working of such institutions, if the public will entrust the Commissioners in Lunacy to extend their interpretation of the meaning of unsoundness of mind justifying detention, and to act in the direction asked for by the Scotch Commissioners; although it may well be that some more stringent provisions against errors in judgment on the part of medical men ought to be provided in the case of insane drunkards than now exist with regard to the more common forms of insanity; and with the elimination of all private interests, such provision which would satisfy the public mind that no injustice was being done, would not be difficult to make. After all, the need of shutting up an insane drunkard, and the right time to restore his liberty, are matters of medical judgment, and of that alone; and if this be given with unquestionable regard to the interests of the patient and of the public, and with no possible deviation towards self-interest, the public will not be keen to distrust the judgment of medical men within their proper sphere.

CONTENTS.

INTRODUCTORY.

Inquiries, &c., respecting the operation of, Inebriate Asylums in America, being a Speech made before the Association of the Medical Superintendents of Asylums in America, at the Annual Meeting, 1875 . . . . *Pages* xxvii—xxx

I.

Speech at the Annual Meeting of the Rugby Temperance Association, 1876, on the Treatment of Habitual Drunkards, and Letter thereupon of Dr. Clouston, Editor of the *Journal of Mental Science* . . . *Pages* 1—40

II.

Letter on the distinction between Habitual Drunkenness as a Vice, and Insane Drunkenness as a Disease.—*Journal of Mental Science*

Letter on the Treatment of Vicious Drunkards.—*Journal of Mental Science* . . . . *Pages* 41—51

III.

On Habitual Drunkenness, a Vice, Crime, or Disease, and the Duty of the State. From the *Contemporary Review*, February, 1877 . . . . *Pages* 52—79

IV.

On the Relations of Drunkenness and Insanity. An Address to a Meeting of the Medico-Psychological Association.
Pages 81—91

V.

Letter to the *Times* on Habitual Drunkenness . *Pages* 92—97

VI.

Letter to the *Times* on Dr. Cameron's Bill for the Cure of Habitual Drunkards . . . . *Pages* 98—100

VII.

Second Letter to the *Times* on the same subject *Pages* 101—103

INTRODUCTORY.

From the Transactions of the Association of Medical Superintendents of Asylums in America, 1875.

DR. BUCKNILL.—I very freely agree with the distinction which Dr. Gundry has drawn between drunkenness which is a vice, and drunkenness which is a disease of a kind to be recognised by an alienist physician. The latter is generally found to be hereditary; the parents were insane or drunkards, or if not the parents, then the grandparents were so, for this disease, like other nervous diseases, frequently skips a generation. Moreover it generally has the characteristic of periodicity, and it is also very generally marked by the existence of other indications of mental infirmity, by eccentric opinions, irrational conduct and dissolute habits; and I think that habitual drunkenness of this kind may very safely be recognised by us as a form of mental disease. I do not see how patients suffering from this form of drunkenness can be detained in asylums during periods of sobriety, unless we do it just as we detain other periodic cases of insanity, cases in which the patient is liable to dangerous attacks of mania at long intervals. We detain such patients during the intervals because we do not know when the attack will recur. I do not see why we should not detain the drunken and debauched maniac on the same ground, but these cases are few and not likely to give us much trouble. The cases which are really difficult to deal with are those in which occasional indulgence has passed into a settled habit of intemperance, and in which eventually the individual seems incapable of resisting temptation; and it seems to me that if such cases are proper subjects of restraint at all, they ought never to be placed in our lunatic asylums, and that we, as physicians, ought never to be made their gaolers. On the question whether habitual drunkenness ought to be considered a disease, I am fully in agreement with Dr. Ordronaux, but not quite on the same grounds, that is, because drunkenness is unnatural, whereas disease is natural; other diseases are not in the order of nature, scurvy for instance. It is no more in the order of nature for men to live on salt pork on board ship than it is to drink whisky, yet scurvy is not recognised as a vice. Ordinary habitual drunkenness I think is a vice and not a disease, and how to deal with it is a perplexing and difficult question. Whether by moral means only, I know not, but I came to this country hoping to obtain much light on this question, which has recently been much agitated in my own country, where we have heard that it has been satisfactorily settled here; and we were much encouraged by the information that a system had been adopted in this country which had proved successful in the cure of drunkenness. Dr. Dodge and Dr. Parrish were induced to come to

England for the purpose of giving evidence upon this most important subject, before Mr. Dalrymple's Parliamentary Committee, and in that evidence these experienced gentlemen stated it as a fact, that from thirty-five to thirty-seven per cent. of habitual drunkards, submitted to their treatment, were *absolutely cured. If that were so, the result would seem to be so grand that the expenditure of large sums of money for the establishment of inebriate asylums would seem to be entirely for the public good; not only for the good of the individuals treated therein, but for the sake of society itself by preventing the propagation of drunkenness, and by diminishing the disturbing element in society of which this drunkenness is so frightful and prolific a cause.* If any members of this Association can inform us what is being done at Binghampton, and what has been done at Media and why Media was abolished, we should feel very grateful. The only practical measure which has been taken in my country, and which has undoubtedly done much good, has been the making drunkenness in public places punishable by law. By a clause in a recent statute, the magistrate can inflict a fine upon any man who is found drunk in a public place; if convicted again within a certain time, the fine may be doubled, and if convicted a third time, within a certain limit of time, the offender may be sent to prison without the option of fine. This simple operative enactment has done more to clear our streets of drunkards, and has done more to check the display of this vice, than anything else I know of. In many places it has worked a reform in the outward and public habits of our people.

I have one observation to make which I think

important. It has been assumed that the habitual drunkard is always the ruin of his wife and family, and that this is a suitable and sufficient reason for the treatment by seclusion, to which it is argued he should be subjected. My experience assures me, that to send all habitual drunkards to institutions of this kind, in many instances would be the ruin of their wives and families. What course should be adopted in cases of this kind? The question would be, shall we send this man to the institution, or leave him to support his wife and family? A dear friend of mine, now no more, gave evidence before the Parliamentary Committee of a person he well knew who got drunk every day for fifty years, and yet during those fifty years he accumulated a large fortune which his family enjoyed. It is not of drunkards who ruin their wives and families, but of habitual drunkards who get all the support for their wives and families, about whom I now speak.

HABITUAL DRUNKENNESS.

HABITUAL DRUNKENNESS.

I.

THE RELATIONS OF DRINK AND INSANITY.

AT a recent meeting of the Rugby Temperance Association, the following speech was made by Dr. Bucknill; in reference to which the succeeding correspondence took place between Dr. Bucknill and Dr. Clouston :—

Dr. Bucknill, in seconding the resolution, said the question of temperance was one in which he took great interest ; in fact, no one could fail to do so who had any regard for the welfare of his race or the progress of his country. He had something specially to say upon one point of the resolution, and should therefore pass over the results of drink in brutality, female degradation, and reckless prodigality, and apply himself to it as a cause of disease both in body and mind. It would be difficult, within any reasonable time, for him to give an outline even of his experience as a physician of the insane, with regard to the production of insanity by intoxicating liquors.

It not only produced insanity directly, but by its effects upon other organs which react upon the brain, and by a variety of causes—by domestic brawls and discomfort to which it gave rise—and it also produced insanity to a frightful extent by leaving it as a fearful inheritance to the children of drunkards. In the production of diseases of the body, he feared the common notions of the disease-producing powers of alcohol were too much confined to what was seen in thorough drunkards, in people who abuse drink to such an extent that they frequently became drunk. But physicians know that that was scarcely the greatest evil. A man who never got drunk, who was never perhaps drunk in his life, might yet drink too much every day, and so shorten his life and weaken his health, thereby stealing away that which was the labouring man's best possession, and which too often the wealthy man could not enjoy—the blessing of healthy existence. He had heard the Revd. Mr. Venables speak with emphasis and enthusiasm of the part which members of his profession were taking in the crusade against intemperance, and he wished he could supplement it by saying that the members of his (Dr. Bucknill's) profession were taking a wise, patriotic, and useful part in the attack upon the great vice of our age and country. But he was afraid that just now members of his profession were taking hold of the stick by the wrong end, and were considering drunkenness not as a cause of disease, but as a disease in itself, which to his mind was a very great mistake. If drunkenness was a disease, it was not a vice, and could not be dealt with by education, and repression, and attempts to reform, but must be dealt with—as indeed many of his profession proposed to deal with it—by establishing hospitals for what they called the unfortunate drunkard. They said, " Poor fellow, he can't help it; he must be placed under medical treatment, and have all the comforts and luxuries he wants, until he is cured." That was not his view of the case. He believed drunkenness to be a fruitful cause of disease, but not in itself a disease; and he looked upon inebriate asylums as an unfortunate attempt to coddle drunkenness, and patch up a wide and fruitful social mischief. Last year he was in America, and took a great interest in visiting the institutions for the promotion of sobriety. He might mention that at the great Centenary he was in Boston, when a crowd of perhaps 150,000 persons went to Concord and Lexington, very fairly to congratulate themselves on the victories their grandfathers won over ours. He mixed with the crowd, and must say they were very disorderly—the police had to make themselves scarce —but he did not see, the whole of the day, in that vast crowd, one man the worse for liquor. He visited many of the American inebriate asylums, and he came to the conclusion that the gentlemen confined in them were generally rather proud of their position, and felt themselves interesting subjects of inquiry. As far as he could observe, they were there under a very lazy and shameful pretence of curing a disease which did not exist, by remedies which were not applied. They had only to walk outside the walls of the institution to the nearest liquor-shop, and get as much liquor as they chose to buy, and they could take liquor into the asylum with them. A friend told him that he went into the inebriate asylum on Ward

Island, for New York, and visited the rooms of four of these unfortunate inebriates, every one of whom was enabled to offer him a choice of spirits. He was not surprised to hear that there was not a very friendly feeling in America between the teetotallers and the supporters of these inebriate asylums. On the previous day he received a Report of the American Association for the Cure of Inebriates, and in that he found a letter from Mr. Carsten Holthouse, a physician to a private institution for inebriates in London, who said, with reference to the relations which exist in this country between teetotallers and the promoters of these asylums :—

> "As regards the bearing of the temperance world generally towards the undertaking—it is not unfriendly ; the more moderate abstainers are decidedly favourable : while the prohibitionists only say, 'You are beginning at the wrong end—providing for the manufactured article, instead of putting a stop to the manufacture.' This section of the temperance people forms, however, but a small portion of the community in this country, and I feel confident that Sir Wilfrid Lawson will never get his Permissive Bill carried in the present generation, and I am still more sure that if he succeeded, it would fail in its object and be evaded in every possible way."

Dr. Bucknill continued, that if the teetotallers were friendly towards Mr. Holthouse, their friendship did not seem to be warmly reciprocated. These gentlemen were urging very constantly and persistently on the Legislature a change in the law which would enable doctors to treat drunkards as poor diseased people—not as he would deal with them, as vicious people, to be repressed and reformed ; or to deal with the question as a great social one, upon which the lines of our educational system should be very much directed. He very earnestly hoped that the Rugby

Association, and the great one to which it was allied, would set their faces against the view of drunkenness as a disease. Habitual drunkenness is not a disease, though it causes all manner of diseases; but in itself it is a vice, and ought to be treated as a vice. The habitual drunkard is a man who likes to drink whenever he can, and who can drink whenever he likes.

ROYAL ASYLUM, MORNINGSIDE, EDINBURGH,
20<i>th April</i>, 1876.

MY DEAR DR. BUCKNILL,—Many thanks for your kindness in sending me the newspaper containing your speech on Intemperance. I confess I was startled at the heresies you express on the question. It seemed as if you were pulling down one of the pillars of our temple.

So far as our case-books here reveal the facts, the following are the answers to the inquiries contained in your note :—

1. Intemprance is the "assigned cause" in 13 per cent. of our admissions here (112 in 878 of all classes during 1873, 1874, and 1875). But of these 878 cases, 310 were put down "unknown," under the head of "causation." If that number is taken off, it leaves 568 with assigned causes for their malady, 112, or 20 per cent. of whom were caused by intemperate habits. But these "unknown" may mean, either that nothing was known of the history of the case, or that the history being known, the cause of the insanity was unknown; in fact, there was no cause to be assigned. In the latter class of cases it was known that intemperance was not the cause, and therefore they ought not to be taken off the whole number, and the

percentage of cases caused by intemperance would not be as great as 20 per cent.

We are as careful as possible about getting the histories of our cases here, but as you well know there are, from various reasons, among such a crowd of admissions as we have here (over 300 a-year) many cases where our information is false, or imperfect, or wanting altogether.

2. I have gone over the last cases admitted here, until I got 100 said to be caused by intemperance. The following are the heads I put them under, and the numbers under each head:—

a. Heredity to insanity	. . . . . .	21
b. Heredity to intemperance	. . . .	6
c. Previous attacks of insanity	. .	23
d. Other bodily causes also present	. .	19
e. Mental ditto ditto		5
f. Cases purely alcoholic	. .	40*

The numbers under *b* are not reliable, questions not having been put on this point in regard to many of the cases. I may say that I knew all these cases myself, so that there is otherwise a fair approach to accuracy in the numbers.

It seems to me, however, that the existence of heredity, or previous attacks, &c., does not much affect the question of intemperance causing mental disease. But for an original instability of brain function of some sort, it would take powerful causes of any kind to produce insanity, and I fancy few asylums would be needed—or few prisons either, for that matter.

* The total of 114 results from the fact that in some of the cases more than one "cause" was assigned, *e.g.* previous attacks and heredity.

If I might be pardoned for presuming to criticise your views, I would say that in the first place you did not fairly represent the opinions of the medical profession when you told your Rugby audience that we all were considering drunkenness not as a cause of disease, but as a disease itself. I don't know any medical man who considers all drunkenness to be a disease, or the result of disease. Most of us do consider that there is a certain kind of drunkenness which is a disease, and not merely a vice. I think you imply that this vice is hereditary, and that it is disease-producing. I confess I cannot myself in all cases distinguish what is vice and what is disease in my drunkard patients, any more than in many of my other insane patients. There seems to be much truth in the idea that disease, its seeds and potentiality, is the vice and sin of the body in many cases, and that the real moral vice and sin are, in those cases, its result and expression. I cannot see that our considering drunkenness as a disease in certain cases should in any way tend to the disuse of attempts to stop and cure it by "education, repression, and attempts to reform." No one says that it is a disease which was always an actuality. It was in all cases at one time of life a mere potentiality, requiring many circumstances to bring it into being. Your measures tend to prevent this, and no sensible man would say that they are the least important. But when the evil germs have grown, is there not room, is there not necessity then, for the disease-theory and the disease-treatment? Can any one deny that all the "repression and attempts to reform" in the world will fail to prevent the neurotic drunkard, whose drunken father was insane, from drinking himself to death, so long as he passes gin-shops every day with money in his pocket? Can any medico-psychologist say that the inhibitory power of such a man over his desires and cravings is as great as that of the average sane man? or that these desires and cravings are not morbid and abnormal both in their strength and direction? Is not the utter and blind disregard of consequences itself a sign of disease, and strictly analogous to the madman's conduct? Is not the loss of inhibitory power over the appetites as great in those cases as over the muscles in chronic alcoholism, and from the same cause, viz., weakening of the controlling powers of the higher brain-centres by alcoholic poisoning?

I so far agree with your views in the practical treatment of all such cases, that along with removing temptations to drinking, I always tell the patient (the sinner—I beg your pardon), that except he wishes to be cured, and tries his best to be cured, no power on earth will cure him. The fact is, your "vice" is always present along with my "disease." I yield that point; but I object to your ousting my disease-theory from the case altogether! I don't see that the practice of American inebriate institutions should make us ignore the facts of nature. It is but natural that the first attempt to deal with this most intractable vice-disease should be uncertain in its result. My notion is much more in the direction of setting up Botany Bays for them, where a change of climate and life would combine with the absence of temptation and with hard work in the open air to alter their morbid constitutions. Then you can't deny that half of them are fools from the beginning, and the other half become fools by their indulgences. They are usually (I mean my diseased drunkards) facile, sensual, irresolute liars, devoid of the rudiments of conscience, self-control, or true affection.

I am, my dear Dr. Bucknill,
Yours very faithfully,
T. S. CLOUSTON.

Dr. Bucknill, F.R.S.

HILLMORTON HALL, RUGBY,
April 27*th*, 1876.

MY DEAR DR. CLOUSTON,—Your welcome letter has been food for much thought, but if I do not sit down to answer it until I have found definite answers to some of the questions in it, it will be a long time before you get an answer.

First let me thank you for so kindly taking so much .trouble to answer my questions about the Statistics of Insanity. I think I will save all I have to say on that subject for the present, and begin with answering, as well as I may, your very fair and weighty criticism on the opinions I expressed at Rugby about habitual drunkenness.

And, first, let me say that those opinions were expressed in an unprepared speech made to a popular audience, upon which I desired to impress a broad conviction. On a different occasion I might have taken greater care to define my position. I do not wish to excuse myself for anything that I did say, but to give a reason why I did not enter into nice distinctions.

Really I think our opinions differ very little, as we might expect, looking, as we do, at the same class of phenomena from the same physiological point of view. I use the word physiological in preference to the word materialistic, which conveys a false impression, if not an imputation.

There is one, and only one, point of fact upon which perhaps we differ—namely, the opinions which have been put forward by medical men on the nature of drunkenness. If you will read Peddie's and Bodington's papers on the subject (read last August before the British Medical Association, at Edinburgh,) you will, I think, see that I was justified in my statement. Dr. Bodington especially is very precise in his declaration that all habitual drunkenness is a disease, and that there are not two kinds of habitual drunkenness, but that "the cases are, one and all, cases of dipsomania, of irresistible, uncontrollable, morbid impulse to drink stimulants." The American Association for the Cure of Inebriates, composed of the Superintendents of Inebriate Asylums, at their first meeting issued a *Declaration* of principle, in which the prime article of faith announced was that " Intemperance is a Disease;" and at all their subsequent meetings, the papers read appear to have been mainly directed to the support of this dogma. And all I have said and written on the subject has been aimed at the mischief which I thought likely to rise from this unqualified opinion. I never supposed that you, or indeed any man able to bring a practised habit of thoughtful consideration upon a large observation of vice and mental disease, could adopt such an opinion without wide reserves and exceptions; but such a man with his appreciation of quantitative and qualitative truth is not likely to appear as an agitator for a great change of law of doubtful wisdom upon a platform of disputed fact.

I think there is very little difference of opinion

between us, if any. I fully recognise the cases you mention—the men who are "facile, sensual, irresolute liars, devoid of the rudiments of conscience, self-control, or true affection," and habitual drunkards withal, as "diseased drunkards." I see that our dear old friend Skae, in the short, but pregnant evidence which he gave before Dalrymple's Committee, maintained the same view [Question 610]. He said, "In speaking of dipsomaniacs there are other symptoms of insanity besides the mere drinking. They are entirely given to lying; you cannot believe a word they say when under the influence of drink, and they will very often entertain a dislike to their friends, which makes them dangerous." I should like to add to this, that, according to my experience, if you are able to watch these cases for some time, you will see short outbreaks of mania not due to drink; and I regard them as a true class of lunatics whose cure is extremely difficult. Perhaps, if there are a sufficient number of them in the country, it would be well that they should be placed under care and treatment in a separate asylum, the management of which might be especially adapted to their peculiarities, and in which they might be detained during a longer period of convalescence than other lunatics, in accordance with a recommendation which I think has been made by the Scotch Commissioners.

But these are not by any means the kind of men I have met with in Inebriate Asylums, nor the kind of men on behalf of whom Dr. Peddie and Dr. Bodington advocate an important change in the law of the land. The Inebriates [what an abominable euphemism this is!] whom I have seen in these asylums have been as devoid of any real signs of mental infirmity as any set of men I ever saw living together in common. And when Dr. Mitchell visited Queensberry Lodge to ascertain whether "any lunatics in the ordinary sense of the word, were there," persons of such a description were not found.

But still more convincing evidence that Inebriates do not correspond with our "diseased drunkards," is to be found in the vaunted results of treatment. Dr. Willard Parker, at the last meeting of the Association for the Cure of Inebriates in the United States, read a paper, the title of which was "Why Inebriate Asylums should be Sustained;" in which he compared the results of treatment in the Binghampton Inebriate Asylum with those obtained in some of the best lunatics asylums in the United States. At Binghampton there are one hundred beds, with an average number of patients of about eighty, and during the year one hundred and thirty-seven patients were discharged *cured ;* while at the New York State Lunatic Asylum there were five hundred and eighty beds, and only one hundred and eighty-two recoveries. You would not expect to obtain such results as the above among diseased drunkards, whatever might be the mode of treatment; and to expect it from the system in vogue in Inebriate Asylums of indolent luxury and *laisser faire* would be in itself almost a sign of imbecility. Either the common run of Inebriates you find in these special asylums are not diseased, or their cure is a philanthropic perversion of fact, or both. Probably both, and when philanthropy sows falsehood broadcast, the furrow produces no crop of annual weeds, but deep rhizomes of untruth, which must be grubbed up with infinite pains and labour.

I think I am perfectly justified in arguing the

Inebriate Asylum question mainly upon the practice of the United States. *The Lancet* has said of one of my statements, " It is absurd to draw from such facts any inference, except of the worthlessness of the statistics of failure which come to us from the other side of the Atlantic." But is it not fair to draw from such facts, also, some inference regarding the statistics of success ? The evidence of the success obtained by the Americans in the cure of drunkenness was the main influence which decided the character of the Report of Mr. Dalrymple's Committee, and the lines of his Bill were laid upon their precedent; and that very Inebriate Asylum for the City of New York, from which I drew the absurd inference, was one of the institutions cited as a model for our imitation. Up to this very moment the men who most loudly demand a change in our law largely affecting the liberty of the subject point to the statistics of success of the American Inebriate Asylums for the cure of drunkenness as their most weighty argument. Moreover, the Superintendents of the American Inebriate Asylums have taken upon themselves a peculiar position as our instructors. They have banded themselves into an association for the propagandism of their dogma that " Intemperance is a Disease ; " and this Association sent a deputation of two of its most prominent members to inform and instruct our legislators respecting the great advantages which we might derive from imitating their proceedings. I think, therefore, that I am perfectly justified in making their practice and their public statements the butt of my criticism.

I feel differently towards the medical men and others who have established Inebriate Asylums in this country. They have had the wisdom or the modesty to refrain from any public demonstration. They have pursued their difficult and unsatisfactory path in comparative silence, and they have received no subsidies from the public purse. They have, without much parade, established private boarding-houses upon temperance principles, in which, no doubt, some benefit is obtained by individuals, and through them by the public.

I feel, therefore, very little disposed to subject them to critical inquiry. When they step forward publicly to teach us the right way to cure the disease of drunkenness, and challenge comparison with the results of treatment in lunatic asylums, perhaps I may have something to say. At present I have only to wish them better success than I fear they have obtained, and to acknowledge the general modesty and credibility of their statements. For instance, in the debate upon Dr. Alfred Carpenter's paper on Dipsomaniacs, read before the Social Science Association in March last, Dr. Ellis is reported to have said that "he had for the last fifteen years kept a private establishment for the reception of persons so diseased, and had had under his charge persons of the highest position—ladies and gentlemen of title; but his experience was that having passed a certain line, they were incurable." But when I see the American inebriate doctors deputed to teach us how to change our laws, vaunting the absolute cure of 34 per cent. of their diseased drunkards, and pushing their creed and their system with an unblushing propagandism, and even challenging our real psychiatry with damaging comparisons; when some of these institutions, moreover, are supported by public funds, and the gentlemen making these statements are public functionaries, then the position seems to be entirely changed, and any one and every one seems to have the right to inquire into the credibility of such statements.

It does not, therefore, seem absurd for me to mention, on the authority of Dr. Macdonald, of the New York City Lunatic Asylum, situate in Ward's Island, that on the occasion of a visit to the City Inebriate Asylum, situate in the same island, he went into the rooms of four of the inmates, and was by each of them offered the choice of spirits.

Nor does it seem absurd for me to state that when I visited the Washington Union for Inebriates at Boston, I was told by Mr. Lawrence, the resident superintendent, that his chief reliance, as a curative measure, was placed in earnest religious exercises, accompanied by temperance songs, supplemented occasionally with pills of cayenne pepper; that his patients had the run of the city, and that he had no means of preventing them from getting drunk out of doors beyond their faithfulness to their word of honour. Nor was I surprised when I met with a man at Binghampton who told me that he had been under treatment at this Washingtonian Home, and that, notwithstanding the religious exercises and the word of honour, he and most of the other patients were in the constant habit of getting whisky at a snug spirit store close to the asylum.

Nor does it seem absurd to me to declare that at the great model Inebriate Asylum at Binghampton belonging to the State of New York, I was assured, not by one patient but by many, that they habitually got as much whisky as they liked by simply walking down to the outskirts of the town, just beyond their own grounds; and that the institution was good for nothing, except as "a place to pick up in"—that is, to recover after a debauch. Nor was I surprised to hear from Dr. Congdon, who has replaced Dr. Dodge as the superintendent of this institution, that he used no medical nor moral treatment. Dr. Gray of Utica, Dr. Burr of Binghampton, and another governor of the institution whose name I forget, heard Dr. Congdon make these admissions to me, and I was told at the time that the impression made upon them was so strong that Dr. Congdon's reign would probably be a short one; which has proved to be the case.

Is it, therefore, absurd to draw the inference that if 34 per cent. of the inmates of such institutions are cured by a residence of a few months, without any real treatment, medical or moral, they have not been the subjects of disease of the brain, nor such patients as we mean when we speak of diseased or insane drunkards? That they may have been drunkards, and that they may have "picked up" and left the institution sober, may perhaps be conceded; but that they have been admitted with one of the most intractable and persistent disorders of the nervous system, and have been cured of it without the use of discipline or treatment, by leading for a brief time a life of indolent luxury, under a cloud of constant tobacco-smoke with cards and billiards, and only ostensible abstinence from whisky, this, if true, would be marvellous.

I must make an exception with regard to the Franklin Home for the *Reform* of Inebriates at Philadelphia under the charge of Dr. Harris. This was the only place I saw in America where honest earnest work was being done, not for the cure but for the reform of drunkards. Dr. Harris repudiates the idea of curing that which is not a disease, and his system is widely different from the no-system which I remarked elsewhere. I will endeavour to give a brief sketch of his method.

He has a set of three single rooms built apart, and which somehow have got the sobriquet of "the barque." When a drunkard—not a patient, mind, but a drunkard—is admitted, generally very drunk, often, indeed, very ill from the effects of a long debauch, Dr. Harris places him in the barque, and keeps him there in bed and in strict seclusion for three days—more, if need be, but three days are usually found to be enough. While there he is at once cut off absolutely from strong drink, not "tapered off," but cut off short. He is also placed upon a limited allowance of water, namely, a pint a day. This is done to prevent vomiting—a frequent ailment with American whisky drinkers—and his strength is carefully built up with strong soups and other nutritious diet. At the end of the three days of solitary confinement in bed he is admitted into the residential part of the institution, to the influences of association with the other inmates, and to earnest exhortations to reform given him by the lay superintendent, and by members of the two committees—one of good men, the other of good women. At the end of a week, if he has picked up pretty well, he is urged to go to work again—not in the institution, but in the City—to face his enemy again, in fact, returning to the institution to sleep. If, as is very often the case, he has drunk himself into poverty and his family into distress, the members of the committees—whom I will not call ladies and gentlemen, for their work is above such terms—help him and his family with money and support, with strenuous help and comfort: and the man must, indeed, be a brute who is callous to such influences.

I will not say that the American is the most reasonable of men, but he is certainly one of the most reasoning, and, therefore, it will appear in no way strange that the inmates of the Franklin Home with whom I conversed manifested a very different tone of feeling to those whom I came across at other institutions. They were penitent and grateful. They leave the institution after a very short probation, and I have no doubt that a very considerable amount of permanent good is effected. Of course there are many relapses, but Dr. Harris discourages repeated re-admissions.

I should like to see institutions on Dr. Harris's principles established at Glasgow, Liverpool, or some other *foci* of spirit drunkenness in our country. They would need no change in the law, for Dr. Harris takes a written consent and indemnity from his drunkards on admission; and if so utterly drunk that they cannot give it, an action for false imprisonment would scarcely lie for their three days' voyage in the barque. It is a reasonable and earnest effort at reformation made without any false pretences, and when it does little good can scarcely do any harm. The drunkards are not coddled in luxurious indolence, nor impressed with the pernicious idea that they are interesting but helpless objects of social and psychological science. They are told the bare truth, and treated, indeed, with the pity due to sinful men by men whom circumstance has only made less sinful; but they are not pampered with false sentiment.

I mark as an important difficulty what you say, that "you cannot in all cases distinguish what is vice and what is disease in your drunken patients, any more than in many other of your insane patients." Still I think you must often be called upon practically to make such a distinction. Most men have some vice, and many men have a prominent vice. When such men, having been insane, have recovered from their insanity, the old vice remains, though the madness has gone, and you have to recognise that which it may perhaps seem rather paradoxical to call a healthy vicious state of mind. But so it is. At least I have found it so, and many a time have had the tough question forced upon me to decide whether pride, or falsehood, or moroseness in convalescence, was a part of the natural character, or the remains of mental disease; and I take it that, even during the disease, it is our difficult but essential duty to distinguish, as far as we can, the two elements of the mixed condition. When a religious and modest woman becomes blasphemous and obscene under child-bearing influences, we do not think her vicious, nor do we attribute all the bad language and misconduct of an insane prostitute to her malady. It is a difficulty which you propound, but it is one with which we are bound to grapple, and does not appear to invalidate the necessity of drawing a broad distinction between vice and disease.

What is that distinction? Where is the *crux?* The *dignus vindice nodus?* From the spiritualistic point of view the answer is easy; but what is the answer from our point of view—the physiological? As a guess at the truth, I would say that vice is a habit of the nervous centres of energizing in an emotional direction, mischievous to the well-being of the individual and of the community, but consistent with healthy nutrition, and not necessarily tending to diminish or destroy the vital activities of the individual. Disease I would define as a condition of some one or more parts of the organism, inherited or acquired, which always involves and implies an abnormal state of the nutrition of those parts, and does necessarily tend, if prolonged and increased, to diminish or destroy the vital activities of the organism. It will be no just objection to this distinction that passion may cause heart disease, and so death; or that a man may carry many local diseases to the end of a long life, terminated by the euthanasia of gradual decay. I think it gives us a fairly just idea of the brain condition in the two states of vice and madness, and supports my view of the way in which we may best prevent or oppose these two different conditions. In the one case by preventing the formation of the habit, or, if it be already formed, by attempting to establish a contrary habit—education and reformation. In the other case, by avoiding the causes of morbid change, or, if the change have already taken place, by endeavouring to re-establish a healthy nutrition—preventive and remedial medicine.

The relation of Drink to Insanity is extremely interesting and important, and so far as I know has never yet been investigated with any degree of thoroughness. In the following remarks, I am far from proposing to enter upon an investigation of this kind, and yet, perhaps, with your help, and that of some other kind friends, one may, without much difficulty, trace the lines of attack.

I use the simple English word Drink, meaning alcoholic drink of every kind; and not that of Drunkenness, because I believe that the habitual use of more alcohol than is consistent with perfect health, although it may never at one time have been used to such excess as to cause absolute intoxication, is a fruitful source of all kinds of disease, more potent, perhaps, than a complete, but rare and exceptional, debauch.

We have no verbal signs which distinguish the habit of drinking from the state of intoxication, as the French have in *ivrognerie* and *ivresse*, but we may agree to use the word Drink to imply alcoholic excess in all its degrees and forms.

Now, it seems to me that Drink may bear two very distinct relationships to the production of Insanity.

It may be the direct cause of insanity as a toxic agent acting on the brain.

It may be one agent among many in the *evolution of insanity*.

If in the old chemical decomposition which delighted our wondering eyes in boyhood, we produce a zinc-tree in a bottle, we get a fairly simple instance of the operation of a direct cause, and we say that the beautiful foliage-like precipitate is the effect of decomposition. But if we compare this simple product of chemical change to a real vegetable growth —to a fern, for instance, which it so much resembles —what a difference is there! The fern is evolved through countless acts of causation which cannot be estimated, and there is no one act of which the most advanced biologist can say—this is its cause.

There are no doubt many cases of insanity caused by alcohol, not quite so simple in their production as the zinc-tree, but still easy enough to understand. The toxic agent, acting on the brain substance, changes its organic composition and deteriorates its function, and we have insanity directly caused by Drink. These cases, I think, are only frequent in populations where heavy spirit-drinking is a common custom; and according to my observations they exhibit the symptoms of dementia rather than those of the more complex forms of aberration.

But what shall we say of those infinitely more difficult cases to understand, one of which is referred to in your able report which I have just received? "When a man with a strong family tendency towards insanity, who has drunk hard previously, is thrown out of employment, and has not therefore sufficient food, and then becomes insane, it is very difficult to tabulate the exact cause of his disease." (P. 10, *Morningside Report*, 1875.)

The distinction of causes into predisposing and exciting, remote and near, physical and mental, &c., will help us to investigate, but will not lead us finally to understand the curious and complex evolution of such a case. Take the drink element, it is predisposing in the early history of the case, exciting later on, it is remote to the insanity, near in the loss of employment, physical always, and yet a part and parcel of the mental state, and the intricate manner in which this red thread runs through the tissue of the life, can never be wholly unravelled. If the previous drink which did not cause insanity had also failed to cause loss of employment, with shame and grief, and semi-starvation, would the mental disease have been evolved?

The drink, as you have stated the case, is the proximate cause of loss of employment, and the remote cause of the insanity; but I think you imply that the drink is continued through all the stages according to the too common history, in which case the estimate of its influence becomes still more embarrassing. The evil begins in the inherited vice of the organism, and as it grows up we get new influences, forming a composition of causes; not applied once for all, but continuing and producing progressive effects, and the history of the evolution comes nearest to that described in the 15th chapter of *Mill's Logic*:—

"The case therefore comes under the principle of a concurrence of causes producing an effect equal to the sum of their separate effects. But as the causes come into play, not all at once, but successively, and as the effect at each instant is the sum of the effects of those causes only, which have come into action up to that instant, the result assumes the form of an ascending series; a succession of sums, each greater than that which preceded it; and we have thus a progressive effect from the continued action of a cause."

It is on these lines, I think, that we may most reasonably hope to get somewhat nearer to the fortress of truth in the more complex cases of the disease which we study.

With regard to Drink we may, perhaps, more conveniently arrange our notions and inquiries under the three following heads :—

1st. Drink causing madness directly.

2nd. Some other influence [as mental strain] causing drink-craving and madness as concomitant results.

3rd. Drink concurring and continuing with other causes, and producing a progressive effect, the end of which is the *evolution* of madness.

I by no means intend to assert that you can always pigeon-hole a concrete case satisfactorily in one or other of these compartments, for there will needs be some doubtful cases, and some hybrids; but the distinction seems founded in nature, and likely to lead to increase of knowledge.

We have much to learn yet, even about the simple direct cases.

I think we must assume, even in the more simple and direct causation of insanity [except, perhaps, from immediate lesions of brain, as by blows or sunstroke], that there is a certain condition of the organization which renders it possible. I entirely concur with what you say that, "But for an original instability of brain function of some sort, it would take powerful causes of any kind to produce insanity," &c. However powerful the causes, many people seem incapable of going mad in the first generation to which such causes are applied, just as I have known three-bottle port bibbers who have never felt a twinge of gout. Without assuming the existence of so marked a state as that which has been called the insane diathesis, we must, I think, premise a certain state of the brain which renders it liable, under efficient causes, to incur those changes of function which we call insanity. This ought, I think, to be considered a predisposing condition, not a predisposing cause; since a cause always produces an effect, but this condition is a barren soil, until the seed of mischief falls upon it. From this point of view, I do not consider heredity a cause, unless it be so strong that it would develop the disease under any circumstances; and even what are called predisposing causes from disease or accident, it would seem right to view rather as conditions suitable to the operation of causes. Thus, a man who has suffered from sun-stroke may be quite rational, if he is exposed to no active cerebral excitement; but to the end of his life a very moderate amount of drink will make him maniacal. The sun-stroke cannot be regarded as the cause of the mania. It has merely been the cause of a certain state of brain, compatible with sanity if the food be simple, but not if it be poisoned. I think the cases of mania à potu from small doses of the toxic agents, which are recorded by Dr. Hayes Newington, in the very interesting paper which you have so kindly sent me, are of this kind. I have myself met with many such cases, most of those I have observed having followed wounds in the head or sun-strokes, or, at least, life in hot countries. They are exceedingly interesting as examples of the brain-condition which I am referring to. I should certainly class the insanity in these cases as caused *directly* by alcohol.

I do not think these cases shift the bearings of the ethical question as you suppose. It cannot make any difference in the morality of the act of drinking, whether it takes a quart or a quartern of whisky to make a man drunk, or one bout instead of many to make him mad. If there be any difference, the greater guilt would seem to be incurred by the greater certainty of mischief, and the man who knows that he will be turned into a maniac by one carouse, is more culpable in his indulgence than those upon whom the evil steals with stealthy and uncertain steps.

I do not understand Dr. Newington to assert that these curious cases of mania à potu, from small doses of alcohol, are characterized by what is called drink-craving, irresistible desire, &c. In my own observations it has not been so, and the *upset* has generally come in some accidental manner. I have never doubted that drink can and does produce insanity directly; and that in some cases a much smaller dose of the poison than usual should be efficient does not seem to change the boundary of vice and disease.

It seems to me that my second pigeon-hole, built elastically as it ought to be, will hold a very considerable number of the cases of insanity roughly referred to drink.

The typical cases are such as one recently mentioned in a letter from Dr. Major of Wakefield, as "a pure case of recurrent mania which has been here five times, in whom one of the first symptoms of the onset of an attack has invariably been a craving for drink, which lasted during the attack, and quite left her when this attack of mania was over."

I take it that most, if not all cases of real oino- or dipso-mania, are of this kind; the symptoms of mental aberration, however, being subject to some variation, being most frequently mild forms of mania, but yet not seldom bearing the mark of emotional depression, but never wholly free from mental disturbance. A sane dipsomaniac is a contradiction in terms.

Here, also we must have a suitable cerebral

condition, not morbid, but *morbific*. A condition compatible with at least temporary health, but susceptible to the influence of exciting causes, which are frequently extremely difficult, and, sometimes, in our present state of knowledge, impossible to recognise. There must be an exciting cause always and invariably for every change of function, for no change can take place without a cause. To say that such and such morbid changes are periodic, is only a verbal veil for our ignorance. It may be that in epilepsy there is a progressive alteration in the balance of certain forces, which needs the thunder-storm of a fit to restore the equilibrium; and in the typical forms of recurrent mania, some process of this kind may be going on during the interval of sanity; but even under this supposition, the final upset of the balance is the exciting cause. In many instances, however, of these recurring diseases, the exciting cause I have no doubt is of a more definite character; for, how shall we otherwise explain the fact that, with great care and quiet, the period is often passed. Very frequently it is a vexation or a passion, or an accidental emotional event of some kind or other. Not unfrequently it is some irregularity in the mechanism of organic life. How little do we know of the small events which may determine such changes! A fatigue, an indigestion, a sexual excess. Anyhow, a positive cause of some kind must operate, or the brain could never pass from a state of healthy into a state of diseased activity, however susceptible it might be, and prone to receive impression. When the exciting cause, whether it be obvious or obscure, has acted, drink and insanity are very frequently the concomitant results. The man drinks because he is insane, and he is the more insane because he drinks. Therefore drink is not a mere symptom of insanity, like incoherence of speech. It is a symptom, but unless interrupted, it reacts as a new cause, and it is not wonderful that undiscerning persons should mistake it for the real and original cause, which has been something quite different.

I am strongly inclined to the opinion that a large proportion of the cases of insanity in our pauper asylums in which the cause of the disease has been returned by the relieving officers as intemperance, are really instances of this kind. Up to the present time the lower class Englishman is pretty sure to resort to drink if he can get it, whenever he acts upon his unrestrained impulses, as when commencing madness blinds him to prudence and propriety. Moreover, when he does give way to drink, it is not in the privacy of his home, but in the glare of the tavern gas; and his intemperance becomes a notorious fact, which is very unlikely to escape the knowledge and attention of the poor-law officials through whose instrumentality he must be protected and relieved.

I know not what may be the case in Scotland, but in those counties of England with which I am best acquainted, I am convinced that if a lunatic of the lower classes has been drinking at all heavily, the relieving officer will be sure to know of it, and will be extremely likely to put down intemperance as the cause of insanity, whether it be so or not. It may be that the Scotch Commissioners are right in thinking that the percentage of insanity caused by intemperance should be calculated upon the admissions in which the cause has been ascertained and stated in the admission papers. But in England I think such a method of reckoning would be misleading. With

THE RELATIONS OF DRINK AND INSANITY. 29

all our etiological knowledge, there are yet many cases of insanity in which we cannot discover the efficient cause of the disease; how many more then in which the imperfectly educated apprehensions of relieving officers would be at fault! Hence this often long list of cases in which no cause has been assigned. But depend upon it, when the pauper lunatic has been drinking heavily, there never is any lack of an assigned cause, whether it be a real cause or only a symptom of his mental state. I do therefore think that the proportion of alcoholic cases admitted into asylums will come nearer to the truth, if compared with the total number of cases admitted, than if calculated upon those only in whom the causes of insanity are supposed to have been ascertained.

A curious and instructive table might be obtained by comparing the percentage of drink cases in the asylums in different parts of the United Kingdom with each other, and with the institutions of foreign countries wherein reliable statistics can be obtained. I have only at hand at the present time very imperfect materials for such a table, but they seem to be sufficient to indicate the extraordinary amount of difference in the part played by drink in the production of insanity in different populations.

As a standard for comparison, let us take Morningside, in which you have been kind enough to ascertain for me that during the last three years 878 cases have been admitted, of which the causes are assigned in 568 instances. In 112 cases intemperance is the assigned cause, being 13 per cent. of the whole admissions, but 20 per cent. of the known causes.

A very fair comparison with Morningside will be the Richmond Asylum in Dublin, in which 53 cases are attributed to "Intemperance and Irregularity of Life," out of a total of 1039, of which number, however, the cause was "not known" in 687 cases—that is, drink was the cause in 5·1 per cent. of all the cases, but in 15 per cent. of the known causes.

In the Friends' Retreat at York, there were 41 admissions and discharges (including deaths), of which 32 had causes assigned; in three instances the cause was intemperance, being 9·4 per cent. in the cause-known cases, and 7·3 per cent. of the whole numbers.

In the Nottingham Hospital for the insane, 34 cases were admitted, discharged, and died, among whom the probable cause was assigned in 29 instances, of which 7 were attributed to intemperance, being 25 per cent. in the cause-assigned cases, and 20·6 per cent. in the whole number. It does not appear whether the 15 cases of heredity are included in the 34, or have to be added to them.

Of the County Asylums, in your own old Asylum for Cumberland, in 142 cases admitted, the causes were unknown in 64, and the cases attributed to intemperance were 6, or 4·2 per cent. on the whole number, and 7·7 per cent. of the known causes.

In the Devon Asylum, of 285 admissions, discharges, and deaths during the year 1875, the cause was ascertained in 238 instances, of which 20 were attributed to "Drink and Dissipation," being 8·9 per cent. of the ascertained causes, and 7 per cent. of the total number.

In the Dorset Asylum, out of 134 cases admitted and discharged, the cause was ascertained in 81 instances, of which 9 were from "Intemperance and Dissipation," being 11·1 per cent. of the ascertained causes, and 6·7 on the whole number.

THE RELATIONS OF DRINK AND INSANITY. 31

In the Warwick Asylum, of 249 cases admitted and discharged (by recovery or death), the cause was ascertained in 206 cases, of which 32 were attributed to intemperance, being 15·5 per cent. on the ascertained causes, and 12·8 on the whole number.

In the Hants Asylum, 275 admissions and discharges contained 233 instances of causes assigned, of which 13 were attributed to intemperance, being 5·57 per cent. of the causes assigned, and 4·73 of the whole number. This proportion seems very small in the county which contains Portsmouth and Southampton.

It will be interesting to compare these percentages with those of American Asylums.

Dr. Kirkbride, in his Report just received, publishes the supposed causes of insanity of the 7,167 cases admitted into the Pennsylvania Hospital since January 1841 ; in 4,301 instances the cause was supposed to be ascertained, and in 637 of these cases it was intemperance (excluding opium and tobacco cases), being 14·78 per cent. in the ascertained causes, and 8·88 per cent. on the total number admitted.

In the State Lunatic Asylum for Pennsylvania, at Harrisburgh, 3,821 patients had been admitted since the opening of the Asylum, of whose insanity, in 2,065 cases, cause was assigned, and in 101 cases this cause was intemperance, being 4·9 per cent. on the cause-known cases, but only 2·64 per cent. on the total of the numbers admitted. A very remarkable difference in the percentage afforded by large numbers in the Pennsylvania Hospital and in the Asylum for the same State. During the last year 178 patients have been admitted into the Pennsylvania State Asylum, of whom 104 had cause assigned, but in only three instances was that cause intemperance, being 2·88 per cent. of the cause-known cases, and only 1·63 per cent. on the numbers admitted.

At the State Lunatic Hospital, Northampton, Massachusetts, 150 patients have been admitted, in whom cause of insanity was assigned in 89 cases; in 10 instances that cause being intemperance, or 11·23 per cent. of the cause-known cases, and 6·6 per cent. on the total number.

In the Hospital for the Insane, Halifax, New Brunswick, the number admitted and discharged in 1875 was 188, in 78 of whom the cause was unknown; in 7 cases the cause assigned was intemperance, being 7 per cent. in the cause-known cases, and 3·2 on the whole number.

In the Minnesota Hospital for the Insane, this year's report states that 1,196 patients have been admitted since the opening of the hospital, of whom in 852 instances the cause was stated. In 57 cases the cause was intemperance, being 6·7 per cent. on the cause-known cases, and 4·8 per cent. on the total admissions.

I have only one more recent report at hand, which gives a Cause Table. It is that for the Criminal Asylum at Broadmoor, and this report differs from all others which I have seen in differentiating the cases attributed to intemperance: 15 cases are attributed to intemperance simply; 2 to intemperance and blow on head; 1 to intemperance and hereditary predisposition; 2 to intemperance and tropical climate; 1 to intemperance and death of husband; 1 to intemperance and domestic troubles; total, 22 drink-caused cases simple or complex out of 70 cases admitted and discharged, of whom 61 were cause-known cases. The percentage of drink-caused cases among criminal lunatics is, as might be expected, very large, namely, 36 per cent. of the cause-known cases, and 31·4 per cent. on the whole number. I have only this day [May 11th] observed the distinction which Dr. Orange has made in his report between the simple and complicated causation of insanity from intemperance, and am much pleased therefore to find that I have the support of his opinions to the need of the troublesome inquiry which I have been asking you and other of my friends who have the means at hand to make into the etiology of insanity from drink. I am sorry that I have not yet received much of this information, which has been kindly promised.

Dr. Duckworth Williams gives the last year's experience of Hayward's Heath for 1875, as follows:—

Males.

Drink simply	8
Ditto operating on hereditary tendency .	1
Ditto operating on pressure of business .	1
Ditto operating on family trouble . .	1
Ditto operating on debauchery	1
	12

Female.

Drink [doubtful]	1

Dr. Parsey gives the experience of the Warwick Asylum on admissions only for 1875, as follows:—

	M.	F.	Total.
Admissions	67	87	154
1. Cases directly the result of the toxic influence of drink upon the brain	5	5	10
2. Indirectly with physical disease or mental trouble	2	2	4
3. With heredity *for insanity*	2	0	2
	9	7	16

In three other female cases without heredity for insanity, one or both the parents were drunkards.

I am inclined to think that heredity from intemperance is a less important factor of insane drunkenness than it is generally supposed to be.

The children of drunkards are grievously exposed to other causes of brain mischief besides heredity, especially to the influences of a turbulent home, and to want of food and proper care during the miserable years of a neglected childhood. It is remarkable that out of 800 idiots admitted into the Earlswood Asylum, Dr. Grabham has only found six instances in which it was stated that intemperance of the parents was the probable cause of the idiocy, and in two of these there was also hereditary insanity. He thinks that the truth in this matter may be often concealed, which is probable enough; but his facts form a striking contrast to those which have long been accepted on the highly respectable authority of Dr. Howe.

However influential in the conduct of life a truth

may be, however wholesome its full force, it is morally wrong and practically mischievous for it to be overstated, which I fear has been done with regard to the heredity of drunkenness. Moreover, if it be admitted that the tendency to drink is transmitted from one generation to another, and that the children's teeth are set on edge because the parents have eaten sour grapes, it does not prove that such an inherited tendency is morbid, for vice also is heritable. As La Bruyère says, " Il y a des vices que nous ne devons à personne, que nous apportons en naissant, et que nous fortifions par l'habitude ; il y en a d'autres que l'on contracte, et que nous sont étrangers."

Magnan's chapter on Dipsomania, in his remarkable work on Alcoholism, seems to support my view, although the eminent author accepts the prevailing theory that dipsomania is a particular form of instinctive monomania, arising, most frequently, from heredity, while alcoholism is a simple state of poisoning, manifesting itself in the same manner in all, even in the brute as in the man.

This distinction will be admitted to be one which ought to be made, if facts exist in nature to support it ; that is, if there be a class of lunatics affected with the instinctive monomania of drunkenness, with complete absence of other signs or indications of unsoundness of mind. It is remarkable, however, that when Magnan produces his evidence, it is destructive of this theoretical distinction. He says—

" Le dipsomane avant de boire se trouve dans les conditions analogues à celles du melancholiaque ; il est triste, inquiet, il dort mal, perd l'appetit, éprouve de l'anxiété précordiale ; *c'est un aliéné ordinaire,* mais après quelques jours d'excès, l'intoxication se produit et le dipsomane se présente avec le délire alcoolique que nous connaissons ; il a hallucinations pénibles, du tremblement, de l'insommie, de l'embarras gastrique, &c., et ce n'est qu'après la disparition des accidents aigus que le diagnostic se complète."

These remarks he supports by an interesting case which had come under his treatment at Saint Anne. A female patient, who, on admission, is pale, agitated, and crying from fear ; she hears assassins who wish to strike her ; she sees at her side the heads of the victims of Pantin ; she believes herself covered with vermin, and shakes her garments ; she hears the voices of her parents, &c., &c. Hands trembling, tongue white, epigastrum painful. No sleep. Hallucinations incessant. The delirium disappeared in five days.

One certainly would say of this patient, "*c'est un aliéné ordinaire.*" But the history of the case given was that for thirty years the woman at certain periods had become sad, interesting herself in nothing, incapable of work, sleeping ill, with no appetite, pain in the stomach increased by the sight of food ; she has an ardent thirst, and drinks wine from the first day, getting it secretly; she drinks until she falls ; she keeps up her drunken state for several days. After the access she reproaches herself, and re-commences her regular and sober mode of life. Formerly the attacks were separated by intervals of fifteen or eighteen months, and at this time drunkenness was the only symptom. More recently the attacks have come on every three or four months, and the alcohol acting more continuously, hallucination and delirium have been developed. She had attempted suicide.

Now, allowing this history to be true, which in one point is an immense assumption, what is there in the case to show that this woman was not a common periodic drunkard, falling very gradually under the dominion of her vice until it resulted in disease, and she became an ordinary lunatic? The one great assumption to which I refer is, that during the long intervals of her attacks she was a sober woman. Let it be remembered that in this country and in France drunkards are allowed by all who know them to be the most inveterate fabricators and deceivers in all matters and questions relating to their vice. In America it is different, and the word of honour of genteel inebriates is implicitly accepted by the confiding physicians who undertake their cure. For my part, I will never trust the word of a drunken man, still less that of a drunken woman, whether palliating their debasement or promising reform. All that M. Magnan records from his own observation about his alcoholic patients I receive with undoubting faith; but of all that he tells of what they have said about themselves I have the deepest mistrust, or unbelief.

Magnan borrows from Trelàt's work another case, which, as he says, makes the distinction between dipsomania and alcholism stand out very clearly. As it is considered a typical case, and affords a good example of the credulous manner in which the drunkard's advocates accept apologetic inventions for sober fact, I shall give it in full :—

Dipsomaniac. Mother and Uncle Dipsomaniacs.

"Madame N——, a person of serious character. She had had during her life many establishments, which have always been wrecked from the same cause. Habitually regular and economical, she was seized from time to time by an *irresistible* access of inebriate monomania, which made her forgetful of everything—of interests, duties, family—and which ended by precipitating her from a position of ample means into one of complete ruin.

"One could not without lively compassion *hear her recital* of the efforts she had made to cure herself of an inclination which has been so fatal to her. When she felt her access coming on she put substances into the wine which she drank, which were best fitted to incite in her disgust at it. It was in vain. *She even mixed excrements in it.* At the same time she spoke insulting words to herself: 'Drink, then, wretch; drink, then, drunkard; drink, villainous woman, forgetful of your first duties, and dishonouring your family.' The passion — the disease — was always stronger than the reproaches which she addressed to herself, or the disgust which she tried to produce. In the last years of her life she was operated on, with success, for a strangulated hernia, and died afterwards of disease of the heart."

I am inclined rather to feel lively compassion for M. Trelât that he has become the historian of such a creature than for Madame N——, though I wonder somewhat that an experienced alienist did not see that if Madame N—— had actually mixed excrement in her drink, she was probably quite insane. If she did this thing without the intention to deceive, she was mad; if she did not do it, she was merely false. Of course one cannot tell from the history which of the two it was; but I think you or I should have ascertained without much difficulty if we could have had the woman under observation.

By being mad I do not intend to imply being in a state of *monomanie ébrieuse*, or the moral insanity of drink, but real aberration of mind, with appropriate intellectual and emotional and physical symptoms, the being *un aliéné ordinaire*, in fact, as M. Magnan puts it. *My position is briefly this—that what is called Dipsomania is either a vice leading to disease in the ordinary pathological sequence; or it is an actual and recognisable form of disease of the brain, with evidence of its existence more cogent than the mere desire for drink.*

With regard to the *irresistible* nature of the propensity which is supposed to prove its morbid origin and to mark it as a moral insanity, it is somewhat strange that the same quality has not yet been attributed to the opium-craving, with which it is most strictly cognate. One would say that the desire for his drug in the opium-eater is far more intense than the craving of the drunkard for his dram, and that his sufferings are keener if the desire be not gratified; and yet so far as I know, Opiomania has not yet been invented as a new form of moral insanity. If there be such a form of insanity, it has been overlooked, in a manner one would not expect in the recent and most interesting paper on Opiophagism from the learned pen of late Commissioner Browne. Tobacco craving also is bad enough when an inebriate smoker has his pipe put out by medical ordinance; and I can answer for it that snuff-craving is no joke under the same circumstances. But these must be trifles compared with opium-craving, which, however, we know to be not irresistible even in its utmost intensity. Neither is drink-craving, if the motive for resistance be greater than the motive for indulgence. Bowring's story in "Bentham" is not so bad on this point—"A countryman who had hurt his eyes by drinking went to a celebrated oculist for advice. He found him at table with a bottle of wine before him. 'You must leave off drinking,' said the oculist. 'How so?' says the countryman. 'You, don't; and yet methinks your own eyes are none of the best.' 'That's very true, friend,' replied the oculist, 'but you are to know I love my bottle better than my eyes.'"

The letter which I sat down to write to you, in answer to your interesting criticism on my little casual speech, has spun itself out into an article which I hope will be acceptable for the pages of the Journal which you so ably edit; and if so, perhaps you will allow it to retain its epistolary form, which must be my apology for the freedom of style which I have permitted myself to use.

   Believe me to remain,
    Very sincerely yours,
     JOHN CHARLES BUCKNILL.

Dr. Clouston.

II.

DR. BUCKNILL ON DRUNKARDS.

To the EDITORS *of the* "JOURNAL OF MENTAL SCIENCE."

SIRS,—

Yesterday I received a printed letter from Dr. Peddie, addressed to me, purporting to be for publication in your Journal, and I natually thought that I owed the sight of this letter, before actual publication, to his courtesy; but this morning I learn from the printer that this letter was sent to me in error. It can, therefore, scarcely surprise Dr. Peddie that, under these circumstances, I prefer to reply to his attack in a letter to yourselves.

In the friendly discussion which I have recently had with one of you on "The Relations of Drink and Insanity," I said—"If you will read Peddie's and Bodington's papers on the subject [read last August before the British Medical Association at Edinburgh] you will, I think, see that I was justified in my statement." That is to say, in the statement that "members of our profession were considering drunkenness, not as a cause of disease, but as a disease in itself." Little did I expect that this reference would have brought upon my head the accusations from Dr. Peddie :—

First—That I have mis-stated and mis-represented his opinions about insane drinkers ;

Secondly—That I have ignored them ;

Thirdly—That I have not read them ; accusations inconsistent with each other, and reminding one of the old pleadings which are now happily abolished, even in the casuistry of the law. It would help me if I knew which count of the indictment contained the real offence, because then, perchance, I might be able to remove or atone for it. To a gentleman who, according to his own statement, has given more thoughtful consideration to these matters "than any other man in the profession," "the felt injustice of having his opinions ignored" might possibly be capable of wounding his self-esteem. Let me hope that the opportunity which he has seized of placing one side of his opinions before your readers in lengthy quotations from his writings, and the further publication of the other side of his opinions which I must ask you to permit me to quote, will induce him to condone this part of my offence, which, I can furthur assure him, was committed most unwittingly. But if I have ignored his "sentiments" about dipsomaniacs how could I have mis-stated them? That is a thing which no man can understand, unless his "*brain-plasms*" can unravel a mystery.

To the third count I must distinctly plead not guilty. Dr. Peddie says—"I cannot believe that you have read a sentence of the paper referred to;" but the real truth is that, before I wrote my letter to you, I had read his paper through several times, in the earnest effort to understand it.

Dr. Bodington—with whose wrath I am also threatened, but of which I am not much afraid, seeing that he leaves one in no doubt about what he means, and, if we differ, as we certainly do, the battle we shall have to fight will be about facts and their interpretation, and not about "sentiments"—Dr. Bodington says—"The confusion between drunkenness as a disease, and drunkenness as a vice, must be cleared up. For my part, I look upon habitual drunkenness as a disease, and I would boldly call it dipsomania. It is in its character as a disease that we physicians are entitled to deal with it. I would sink the notion of its being a mere vicious propensity. When fully developed there are not two kinds of habitual drunkenness. The cases are, one and all, cases of dipsomania, of irresistible, uncontrollable, morbid impulse to drink stimulants."

That, without doubt, is a sentence entirely devoid of "hair-splitting distinctions." No two sides of the same shield there painted different colours; or dark cloud with a silver lining.

Dr. Peddie has quoted a large portion of his paper (though it was already accessible enough in the pages of the "British Medical Journal") to prove that " my [his] *sentiments* have ever been such as appear to accord with those you have quoted as Dr. Clouston's and, consequently, that we are *all three in truth agreed* as to the persons who may be styled dipsomaniacs!" But, if so, where is the need for dispute?

In point of fact we are by no means agreed, for the real gist and purpose of Dr. Peddie's paper turns upon his 4th class, namely, those who acquire "the propensity to intemperance" through a course of vicious indulgence in stimulants. About maniacal and delirious cases he admits there can be no doubt; they are proper subjects for a hospital or an asylum, but it is for the "unfortunate individuals who are so perplexing to themselves and to society," and who cannot be placed in hospitals and asylums because they manifest no symptoms of disease of mind or body beyond the propensity to intemperance, it is for them he advocates a change of the law under which they can be profitably kept in a new kind of sponging-house, or private gaol for drunkards instead of for debtors. When Dr. Peddie gave evidence before Dalrymple's Select Committee some of the members tormented him into a precise statement of his sentiments, and here they are:—

"Question 1016. Dr. Playfair—You say that you would take a man and put him into forced detention; under what condition would you do that?—When a man could no longer control himself from the habit of intemperance, I would then consider him in a condition of unsound mind and requiring to be cared for.

1017. Even if he was only injurious to himself, and not immediately injurious to the public?—Yes, I think that we should do something more than provide against injury to the public; I think we have a duty as citizens and fellow-creatures to one who will not take care of himself.

1059. Mr. W. H. Gladstone—Do you not foresee great difficulty in determining when a man may be said to have lost his power of self-control?—No, I should not feel any difficulty; I think that it is a matter of medical diagnosis. There is not more difficulty in regard to the habitual drunkard than there is difficulty in regard to insanity of other forms;

medical men have constantly cases of insanity brought before them, and the question in each case is whether or not such an individual is a proper subject for control in an asylum for curative treatment.

1060. Then do you think that a man who, when sober, is in complete possession of all his faculties, may still be said to have lost all self-control?—We know very well that we should be able to distinguish in that case his danger by the supposition that if drink was placed in his way the next day, or that very evening, he could not resist it, and that if he once tasted it he would go on from bad to worse; a craving would be set up of which there has been a frequent opportunity of judging before, and that he would go deeper and deeper into the mire.

1061. Do you think that the impulse to drink, in a case like that, is different from other vicious impulses, such as, for instance, an impulse for gambling?—Yes, I think that the impulse is quite different.

1062. It partakes more of the nature of an external disease, like fever, which comes upon a person?—I consider it greatly in the nature of an internal disease; there is also alcoholic influence and some kind of change upon the state of the brain thus affecting its operations.

1063. But it is analogous to an ordinary disease?— It is analogous to an ordinary disease."

Surely I have a somewhat better right than Dr. Peddie to complain that my opinions about insane drunkards have been misrepresented when they are declared to be in complete agreement with those of a writer who maintains that a man may be an insane drunkard "who, when sober, is in complete possession of his faculties."

Dr. Peddie "would not feel any difficulty in determining when a man has lost his self-control." "It is a matter of medical diagnosis. There is not more difficulty in regard to the habitual drunkard than there is in insanity of other forms."

But is it not somewhat inconsistent with this avowal that Dr. Peddie should now insist that this diagnosis cannot be adequately made by men who have the greatest knowledge of insanity of other forms, because "specialists in lunacy cannot come in contact with many cases of genuine dipsomania? They can only see a fraction of such cases as come under the notice of physicians in ordinary practice."

As specialists in lunacy know so little about these genuine cases of insane drunkenness, it is not altogether unreasonable that they should be warned off this domain of the physician in general practice. Consequently "the cure of dipsomoniacs" must not have "any connection with lunacy arrangements." "Lunatic Asylums are not adapted for the reformatory treatment in such cases."

If these are the cases which, when sober, are in complete possession of their faculties, specialists in lunacy will not perhaps act unwisely if they resign the honour of their treatment to those who understand it so much better; but Dalrymple's Committee had other views as to the knowledge of such specialists in lunacy, or they would not have called before them as witnesses such men as Drs. Crichton-Browne, Skae, Mitchell, Nugent, Boyd, and Mould, who contributed for their information many important elements of diagnosis which we do not find in Dr. Peddie's writings, notwithstanding that he has thoughtfully considered this matter for such a very long time. I am sure that these eminent specialists in giving their evidence desired no more to give a specialist colour to the facts garnered by their vast experience, than in writing my letter to you I wished "to raise a cloud of psychological dust to defeat or discourage a highly philanthropic movement." The movement may be highly philanthropic, but there is another kind of dust, namely, gold dust, which seems to have some influence in urging it on, for to quote Dr. Peddie once more :—

"In order to call into existence houses or institutions such as would be suitable for the upper and middle classes of society, a law to empower restraint and detention is manifestly essential. A few such institutions on a small scale have existed in Scotland, but have laboured under most discouraging difficulties from want of authority to receive and retain a sufficient number of inmates, and for a sufficient length of time, to become remunerative. This has stood in the way of liberal investment for suitable premises, ground furnishings, staff of service, etc. Thus the important essentials for efficient treatment have been necessarily defective; and the result is, that the care of a very small fraction only of insane drinkers has been undertaken, and cure somewhat rare.

"The inmates, with partially restored sanity from enforced deprivation of stimulants, become restless, and knowing that they cannot be detained legally, demand liberty, and take leave, or else work on the minds of friends or guardians by entreaties or threats, and get it. If, however, the State will sanction, under proper checks, both voluntary admissions and compulsory commitments, in cases of genuine dipsomania, permitting prolonged detentions, until real benefit is derived, a sufficient number of Homes or Retreats, or by whatever name they may be called, for the cure of persons in the upper and middle classes, would certainly spring up, both through private enterprise and the efforts of companies or associations, formed for the purpose, somewhat similar, indeed, to many existing and thriving lunatic retreats and asylums, affording accommodation and means of treatment very different in efficiency from those inebriate institutions which have, in times past, struggled under cramping difficulties. Now, into such houses as these, many unfortunate persons would enter voluntarily, as they do in some of the American inebriate institutions, knowing that, if they did not thus surrender themselves for treatment, they would be compulsorily committed; and then, when they are under control, the law, as I have already hinted, could prolong it for such a time as might be deemed necessary to accomplish the humane ends in view."

Alas! alas! that it should all come to this! This highly philanthropic movement! These humane ends in view!

When I think, sir, of what the evil of strong drink really is among the lower classes in some parts of your country and of mine; when I think of what I saw in company with Sheriff Dickson in the drink-haunts of Glasgow on the night of Saturday the 27th of May last; when I think of the crowds of men and women, many of them infant-laden, whom I there saw steeped in the bestiality of drink, it makes me right angry with these philanthropic *fribbles*, who, with eyes averted from the drunken and debased populace, fondle the subject of the casual rich man's drunkenness, with dainty considerations of how he is to be placed in a golden cage, " pleasing his palate in the way of good culinary arrangements," and his captivity made profitable.

Let Dr. Peddie carefully examine the wynds of Glasgow, their drink-shops, lodging-houses, and policecells, on a Saturday night, and he will afterwards perhaps not think it so easy to perfume hell with rosewater.

As I said in the speech which has led to this discussion, some members of our profession are misdirecting the attention of the public in this matter. By the noise of their philanthropic drum, they would lead us, by false alarms, from the real field of battle. They dally with the tarnished fringe of drunken society, while its broad expanse is a funeral pall to myriads of lowly victims ; and Dalrymple's Committee, with its foregone conclusion, unwittingly established the dreadful fact of alcoholic eremacausis in our swarming cities, and concluded by recommending a most dangerous and unconstitutional change in the law for the supposed benefit of those classes of society in which a drunkard is becoming a somewhat rare specimen of a decaying and dishonoured vice. They made out the charge fully against the common folk, at least in certain localities, and they directed the main force of their proposed remedy against the stragglers and backsliders of the sober classes. They would scarify the field with a chain harrow when it stands in urgent need of deep draining and subsoiling.

Dr. Peddie, to give him his just due, has not altogether passed on the other side from the drunken crowd, for in his evidence before the Committee he proposed the establishment for the whole of Scotland of four public inebriate asylums, each to contain forty patients of the working classes. They were to be model institutions. He admitted that all four would not contain the habitual drunkards of Edinburgh alone, and, indeed, he may any day find nearly twice as many of the gentle sex in Queensberry House. But it was honourable to him, considering the example of some of his co-agitators, that he allowed his mind to dwell for a moment upon the treatment of drunkards who cannot pay. Public provision for the treatment of 160 working-class drunkards for the whole of Scotland, and for the idle class drunkards as many private houses of detention as the law of profitable investment, aided by that of "compulsory arrest," may develop, reminds one of the proportions

E

of Falstaff's bread and sack, in the relative regard for
the class which represents the staff of life, and that
which drinks the wine of its wealth and luxury.

Dr. Peddie also suggests [see Appendix of *Report
on Drunkards*, p. 187,] that "the *pauper* class of
drunkards should be taken care of in the separate
wards of a poor-house," and that "the *criminal*
drunkard class should be accommodated in wards or
separate houses connected with our chief prisons."
"By these arrangements," he thinks, "the unhappy
individuals would have more chance of benefit from a
distinct and more *attractive* system of treatment."

In these separate wards, to be called Reformatories,
work is "to be made both agreeable and profitable
by a *system of rewards and benefits.*" For the rich
drunkard the loss of liberty is to be sweetened by
manifold attractions, of which "not the least would
be *perfection in the culinary department,*" and "such
new and *relishable* enjoyments as might counteract or
take the place of craving for alcoholic stimulants."

All this, indeed, is philanthropy and not science,
not even social science. Perhaps it is not even "non-
professional common sense," for we should all wish to
be Inebriates that we might enjoy ourselves under
the protection of Dr. Peddie's wing, and he might
become the only sober man left in the land. What a
position, *Sa.ius, Solus, Sobrius, Rex ebriorum!* Only
there would be no bread-winners and rate-payers left
to support the drunkards—I beg their pardon—the
Inebriates. But even this bit of a difficulty might
possibly be averted by Dr. Peddie's ingenious suggestion that Inebriates may be allowed to carry on their
work or business, their wages or profits being taken
away from them, and "so leaving them *free to earn*

but not free to spend;" a suggestion which indicates a knowledge of human nature more profound than even " non-professional common-sense " can fairly reach.

I am extremely sorry to have caused Dr. Peddie "the felt injustice of having his opinions ignored." The truth is, that when I wrote to you on "The Relations of Drink and Insanity," I was entirely preoccupied by the consideration of the question, and had no thought, purpose, or notion of giving Dr. Peddie the slightest offence. Should this letter also not please him, I must insist that it is no fault of mine, seeing that I have been constrained by him to introduce, most unwillingly, into the discussion of a scientific question, matters which may seem to have a somewhat personal bearing. But, when a man of Dr. Peddie's eminence asserts that in such a discussion you are unjust if you ignore " my " opinions, one is compelled, as it were, to stand and deliver one's opinions upon his opinions, whatever they may be. I very much wish that mine could have been more in agreement with them.

I am your obedient servant,

JOHN CHARLES BUCKNILL.

39, WIMPOLE STREET, *August 24th*, 1876.

III.

HABITUAL DRUNKENNESS: A VICE, CRIME, OR DISEASE?

"Who shall be the Rectors of our daily Rioting? and what shall be done to inhibit the multitudes that frequent those houses where drunkenness is sold and harboured?" are questions put by Milton in his "Areopagitica," which have since so greatly swollen in magnitude that they can no longer be answered in his manner, nor the encouragement of sobriety be left as he would have it, "without particular law or prescription, wholly to the demeanour of every grown man." The growth of this "national corruption" has been so great since his time that only despair of amendment could seat us on the stool of inaction with folded hands, acknowledging with him that "these things will be and must be;" or if this be so, we are compelled to feel that, how these things "shall be least hurtful, how least enticing, herein consists the grave and governing wisdom of a State."

It is strange that Milton should have drawn so much of his argument for the liberty of unlicensed printing from the supposed liberty of unlicensed drunkenness, seeing that the only law against drunkenness, pure and simple, which we still have, has been on the statute-book from the reign of James I. This statute, which imposes a fine of five shillings, neither more nor less, upon a drunken person, does certainly constitute the act of becoming drunk, however privately or quietly it may be done, into an offence against the law of the land. But notwithstanding the existence and the occasional application of this old statute, most modern writers on the philosophy of legislation have maintained that private drunkenness ought not to be the subject of legal repression. No one has written more forcibly on this point than Bentham himself, who in his "Principles of Morals and Legislation" declares that the primary mischief of drunkenness is "private and self-regarding," and that its secondary mischief of example will not often amount to a danger worthy of notice.

"With what degree of success," he asks, "would a legislator go about to extirpate drunkenness and fornication by dint of legal punishment? Not all the tortures which ingenuity could invent would compass it ; and before he had made any progress worth regarding, such a mass of evil would be produced by the punishment as would exceed a thousandfold the utmost possible mischief of the offence. The great difficulty would be in the procuring evidence, an object which could not be attempted with any probability of success, without spreading dismay through every family, tearing the bonds of sympathy asunder, and rooting out the influence of all the social motives. All he can do, then, against offences of this nature with any prospect of advantage in the way of direct legislation, is to subject them, in cases of notoriety, to a slight censure, so as thereby to cover them with a slight shade of artificial disrepute."

In his " Essay on Liberty," John Stuart Mill argues this point still more keenly and thoroughly, upon the principle that "no person ought to be punished simply for being drunk ;" and he very clearly indicates the conditions and complications which might render punishment for drunkenness justifiable, such as the neglect of public duty, the inability to pay debts or to support and educate a family, and the act of " making himself drunk in a person whom drunkenness excites to do harm to others."

Milton, Bentham, and Mill, then, agree in this, that simple drunkenness ought to be subjected not to the political but to the social sanction, and that "all the legislator can hope to do is to increase the efficacy of private ethics." This to a great extent has been done among the cultured classes by the growth of the sentiment that intemperance is dishonouring and shameful, and among the masses it is being done by the steady increase and influence in this country of temperance societies, which already number three millions of members, affording a large practical and promising answer to Mill's demand for a supplement to the unavoidable imperfections of law, if not in the form of a "powerful police," against this vice, at least in that of a watchful, pervading, and persuasive opinion.

If "no person ought to be punished simply for being drunk," the persistent demand which we hear for a new law under which men may be subjected to long periods of imprisonment because they are in the habit of doing that which ought not to be made an offence is only intelligible on the supposition that the repetition of the vice constitutes a disease. This indeed is the real justification for the proposed enactment which the Parliamentary Committee on Habitual Drunkards of 1872 put forward in their Report. They say :—

"Occasional drunkenness may, and very frequently does, become confirmed and habitual, and soon passes into the condition of a *disease* uncontrollable by the individual, unless indeed some extraneous influence, either punitive or curative, is brought into play."

This proposition, however, that confirmed and habitual drunkenness *passes into* the condition of a disease, is a much more limited and cautious one than that which is propounded by the professional advocates of what are called by the euphonious name of Inebriate Asylums. With them habitual drunkenness is itself a disease, caused in some curious physico-metaphysical manner, by what they call paralysis of the will, and manifested by supposed want of control over the conduct. This opinion was frankly expressed in the paper read by Dr. Bodington before the last annual meeting of the British Medical Association, and it is to be found pretty generally announced in the abundant medical literature on the subject. Dr. Bodington says:—

"The confusion between drunkenness as a *disease* and drunkenness as a *vice* must be cleared up. For my part I look upon all habitual drunkenness as a disease, and I would boldly call it all *dipsomania*. It is in its character as a disease that we as physicians are entitled to deal with it. I would sink the notion of its being a mere vicious propensity. When fully developed there are not two kinds of habitual drunkenness. The cases are one and all cases of *dipsomania*, of irresistible, uncontrollable, *morbid impulse* to drink stimulants."

A still more remarkable instance of the extreme position which has been taken on this question has been afforded in the proceedings of the American Association for the *Cure* of Inebriates. At the first meeting of this Association a declaration was issued in which the dogma was solemnly propounded that "intemperance is a disease," and various papers were subsequently read by Dr. Parrish the president, and others, to explain and maintain this prime article of faith. At the fifth meeting of the Association, Dr. R. P. Harris, the physician to the Franklin *Reformatory* for Inebriates at Philadelphia, was bold or incautious enough to present a report, in which were certain expressions more in conformity with the name of his Institution than with the Creed of his associates. "He regarded drunkenness as a habit, sin, or crime, and did not speak of cases being *cured*, as in a hospital, but *reformed*." The Association denounced this heresy on the ground that the truth of intemperance being a disease was the base of their organization, failing which their very name would be "a fraud upon the community;" and their Publication Committee was instructed "to return the report to its author, with a request that it be modified so as to conform with the Declaration of the Association, with power to publish such parts of it as they may deem proper." The Committee published the *revised report* entire.

America is not quite

"The land where, girt by friends or foes,
A man may speak the thing he will."

The majority there has a power of excommunication much more active than any which pope or priest ever possessed, and he who questions the dogma of disease among doctors must be liable to the penalty. A dogma put at the head of a declaration too! Why one might almost as well question that great dogma with which the constitution itself commences, "Forasmuch as all men are equal!" It is unfortunately a fact that if you write an assertion of this kind in very large letters, it is often accepted by the world without doubt or question, until belief in it becomes habitual and uncontrollable, and according to those marks a kind of disease; so that belief in the morbid nature of intemperance may itself become morbid, if its passionate, inveterate, and uncontrollable nature is equally admitted as the true characteristic of this condition.

Before making any attempt to examine the real nature of habitual intemperance in drink, it will be fair and right to mention, for the purpose of exception, those instances in which drink has produced common bodily disease, and those in which disease has been the occasion of the drunkenness. Cases of the first kind, in which the mental faculties are not involved, are perhaps the heaviest portion of the curse which strong drink lays upon us; but they do not and cannot enter into this argument.

But that strong drink does often cause disease of the nervous system, with disturbance of its mental functions, and also that such diseases of the mind, arising from other causes, do also give rise to the passion for drink, are facts which can admit of no doubt. Every medical man is more or less conversant with them, and medical men who have made mental diseases a special study are well able to recognise them. The history of such cases, their heredity, periodicity, the intermixture of their mental symptoms with other nervous disorders and defects, the peculiarities of the mental symptoms themselves in certain forms of emotional disturbance and of intellectual aberration and decay, are very well known to physicians who have made madness their study. The number of such cases, enormously exaggerated as they often are, is yet large enough in lunatic asylums and in private practice to afford abundance of experience as to the marks of all those conditions which can truly be called insane drunkenness, or drunken insanity; and they may well be left to the observation and treatment of the specialist physicians who understand them.

There is, however, another class of cases, fortunately a very small one, but of extreme difficulty to understand. Perhaps also it is fortunate that ignorance of their nature does not seem much to matter, since the physicians who have met with them pronounce them to be absolutely incurable. I refer to those cases for which the learned terms *oinomania* and *dipsomania* were originally invented before the latter term was filched from science to designate vulgar drunkenness. In a somewhat large experience, I have myself never yet met with an undoubted instance of pure *dipsomania;* and I observe that very few examples are on record in medical literature, and that these are copied by one author from another in a manner which sufficiently testifies their rarity. The evidence, however, of credible observers is perhaps sufficient to establish the fact of their occasional occurrence. Dipsomania is described as a form of moral insanity manifesting itself in a passion for strong drink, not for its own sake, or for the sensations which it produces, but for the gratification of a morbid impulse. As the kleptomaniac steals not for the sake of possessing the thing he steals, and the homicidal maniac destroys life not for the purpose of making any person cease to live, so the dipsomaniac drinks, not because he likes drink, or likes to get drunk, but because he has an uncontrollable and morbid impulse to swallow intoxicating liquor. To constitute a pure case of dipsomania all other manifestations of unsoundness of mind ought to be eliminated, for if they exist the

A VICE, CRIME, OR DISEASE?

case is simply one of insanity with one prominent symptom. All other marks of a vicious disposition ought also to be absent, for if they exist, the dipsomaniac is simply a drunkard able to make superfine excuses. To be up to the pattern, he ought to be thoroughly sane and eminently virtuous. For my own part, I suspect that he deviates from the pattern sometimes in the warp and sometimes in the woof, and sometimes in both. However, I am ready to discuss him whenever he appears again, like the great bustard in Norfolk, and I only mention him here in order that he may not suffer the indignity of being mixed up in an argument about vulgar drunkards who are as common among us as sparrows in the hedgerows.

We come then, by process of exclusion, to simple habitual drunkenness, apart from any other symptom of mental disease than the passionate and irrational indulgence of a noxious habit.

Is this a disease or a vice, and ought we to direct our efforts to its cure or to its reform?

To many people the answer will be very simple, seeing that they will consider disease as a condition of the material, and vice as an affection of the immaterial element of our nature; but such is not my view, and I shall endeavour to consider both terms of the question by the same physiological method. And here it is needful to premise that as there is no such thing as a disease in itself, so there is no such thing as a vice in itself. Doctors have fallen into the habit of realizing and classifying diseases as if they were all composed of various *materies morbi* and were actual things; nay, sometimes, they almost personify them as in the old trope of "the pestilence which walketh in darkness," and moralists have done quite as much for vice. But the idea of a disease, be it ever so specific, can only be a mental creation generalizing diseased bodies, and of a vice the same, generalizing vicious beings. Reality ever attaches to the individual; and here let it attach, say to John Jones, and the question be whether said John Jones, being often drunk, is diseased or vicious?

I think it must be admitted that when said John Jones is actually drunk his organism is in a state of disease, *pro tanto*, as the same must be admitted of him when he has eaten more than he can easily digest, or when he is exhausted by debauchery, or actually suffering disturbance of healthy function from any other sensual excess; for any condition of the organism by which its healthy functions are disturbed, and which, if indefinitely prolonged and increased, would tend to their suspension and even to death, must be called a diseased condition.

The contracted question then comes to be whether John Jones is diseased or vicious while he is still sober, and when he is about to drink the " brewed enchantment," knowing all the effects which it will have upon him?

Now, there are several causes and qualities of this condition which are common to the notion of disease and to that of vice :—

a. The tendency to disease is sometimes hereditary, and so is that to vice.

b. The causes of acquired disease are sometimes small, gradual, and accumulative, and so are those of vice.

c. By continuance and repetition diseased conditions become inveterate, and so do vicious ones, indeed by far the more so; and it would be a happy thing for mankind if the clergy could reform men with the same success which doctors attain in curing them.

d. Disease is cured by removing the cause, *sublatâ causâ tollitur effectus*, and vice is abrogated by the same means; and in both, when the cause returns, the effect is reproduced.

These seem to be the common qualities of the two conditions.

Some of the qualities by which they are differentiated would seem to be:—

a. Disease consists solely and entirely in some change in the organization, which is often known to, and is always thinkable by, the physician. This is a matter of certainty. But it is not known that vice consists in, or is even accompanied by, any such change. Certain vices may produce such a change, as an effect, but such changes are not known to exist as constituent conditions of vice; and if they do exist, they are too obscure for our present means of observation, or even for our power of thinking of them by analogy or imagination.

If it be admitted that both vice and disease are dependent upon states of the organism, a consideration of one fact will suffice to prove that these states are very different. In the early stages of disease, and especially of nervous and mental disease, there may be no changes after death appreciable by our senses or means of examination; but in later periods of such disease such changes are mostly to be found. But in uncomplicated vice, whatever its duration and whenever death may occur, no such causal changes can be observed.

If there be really no difference between vice and disease, all punishment, nay, even any reprobation, is unjust. Physicians should replace the magistrate and the priest, and courts, prisons, and churches be converted into hospitals.

b. The causes of disease are physical, and the last link in the chain of causation—that is to say, the causal condition—is invariably so. Emotion may be the exciting cause, but there is always a physical link intervening, generally in the form of disordered innervation. The cause of vice, on the other hand, is always moral, even where the conditions of the vice are grossly material and sensual.

c. The remedies for disease are mostly physical, and are invariably of a physiological nature. Even "medicine for the mind diseased," in what is called the moral treatment of insanity, is really directed to the material condition of the organ. The remedies for vice are of a different nature, and are mainly directed to elicit opposing desire, to make indulgence more immediately painful, and to influence the judgment.

d. The successful result of remedies, which in one case we call cure and in the other reform, is very different. The cure of disease establishes a healthy condition of the body; the reform of vice establishes a virtuous condition of mind, and, even should the latter be looked upon as a condition of the emotional functions of the brain, the notion of it is different from that of merely physical health.

e. With some exceptions, persons suffering from mental disease are not conscious of their misfortune. "Le premier degré dans l'homme après la raison," says La Bruyère, " ce serait de sentir qu'il l'a perdue ; la folie même est incompatible avec cette connaissance." The vicious man is generally conscious of his vice, which, unless thoroughly brutalized, he regards with shame and remorse. Disease may occasion regret, but not remorse.

f. Few men are diseased, but all are vicious. The one state is an accident to man's nature, the other is an element of it. And, with regard to the vice of drunkenness, there never yet was found a tribe of savages so unsophisticated that they needed more than the opportunity and the first lesson to plunge headlong into the abyss. The first debauch may be the fault of ignorance, as perhaps that of Noah was when "he drank of the wine and was drunken;" but the savages of all climes are habitual drunkards *in posse*. And the wild theory that habitual drunkenness is a disease supposes that the existence of a keg of fire-water will convert a camp of Indians into madmen while even they have not yet tasted it, for while they are sober they will barter their possessions, their liberty, almost life itself, for its enjoyment.

If we examine the condition of a man who has what is called an irresistible and uncontrollable desire to imbibe strong drink by these tests, we cannot fail to see that it is not that of disease, but that of vice.

The passion for drink, however habitual and forcible it may be, is not associated with any physical change in the organism. The indulgence of it produces changes which generally lead the drinker through a life of misery to an early grave, but even this effect is not constant, and before it has taken place the organism of the drunkard may be, and sometimes is, a thoroughly healthy one. If alcohol be a poison, it is not a very rapid one; and the statement which Dr. Forbes Winslow made before the Parliamentary Committee, that "very often with chronic drunkards, on examination after death, if you apply a light to the fluid in the ventricles of the brain it ignites into a flame," seems to need verification. If found to be correct, it points to one use to which a drunkard might be applied : he might be distilled.

But a drunkard is such before he is drunken and in his sober intervals, and we are considering the condition of his brain when it is not "saturated with alcohol;" and this, I say, may be perfectly healthy. Excessive indulgence in other sensual vices—lust, for instance—will sometimes cause organic changes, such as softening of the spinal cord, but while the vice is only part of the character there are no such changes; they are the physical results of physical indulgence. The same is even true of the more purely mental vices, as anger, the physical results of which sometimes lead to morbid changes of the heart.

Then the *cause* of the desire to drink, the *motive* influence, is not physical, but moral. A man drinks because he likes it; he likes the taste of the liquor, and still more the exhilaration which follows, and even the *narcosis* which succeeds. There can be no doubt that drinking, and even drunkenness, is pleasureable to the vast majority of mankind; and if drinking brandy were as harmless as eating ripe fruit, it would be difficult to make it a vice. But the effects on mind and body are such that the present pleasure is grievously outweighed by the subsequent pain. The pain of exhaustion is the first which comes, and drives the man back to the stimulant, weaving the web of habit. The drunkard, like all other vicious men, prefers the present pleasure to the absence of pain in the future, that is, to future pleasure. He is a bad moral calculator, ever running counter to the wise maxim of Seneca, "Sic præsentibus utaris voluptatibus, ut futuris non noceas."

The remedies for *drink craving*, as it has been called, are all moral. Even the inebriate asylum is a moral remedy directed to change the character, not to cure the disease; for if cure only were aimed at, the drunkard would be dismissed in a few days, as soon as he could digest his food and sleep o' nights like sober folk; but the cry is that he must be detained for not less than two years, in order that his character may be changed by absence of temptation. This surely is directed to the moral side of his nature.

All punishment of course is moral; whether it is of that sure and constant form which the drunkard draws upon himself—

"Oh, sir, to wilful men,
The injuries which they themselves procure,
Must be their schoolmasters"—

or the precarious punishment inflicted by the State or by society, its influence is purely moral. But whether by removal of temptation, or by self-inflicted suffering, or by punishment, the result aimed at is always a change in the drunkard's character, not in his health; and this change we call a reform not a cure. That this change is really so rare is an indication of its nature, the eradication of an inveterate habit of mind. The American physicians who gave evidence before Mr. Dalrymple's Committee asserted that they *cured* 34 per cent. of the drunkards who were admitted into their asylums, and it might be so far true, that this proportion of their patients were sober and healthy when they left their asylums. But the Commisioners in Lunacy for Scotland, who possess large opportunities for observation, have come to a very different conclusion as to the frequency of the real reform of a drunkard. These public men of large views and wide experience, who have watched the considerable number of habitual drunkards who place themselves voluntarily in the Scotch asylums, and who also officially visit the Scotch inebriate asylums to see that no really insane persons are detained therein, in their Reports for 1872, 1873, and 1874, stated their opinion that

> "It is possible that prolonged compulsory abstinence from alcoholic liquors may restore to habitual drunkards the power of self-control, and enable them to resist the craving to which when at liberty they succumb. Our experience, however, does not give us much reason to expect this result."

And to this passage in the first of these reports, the following very remarkable addition is made : "*Indeed it would not be easy to point out one single case of permanent and satisfactory reform.*"

The same opinion was expressed to the Parliamentary Committee by Dr. Mitchell, one of these Commissioners, whose wise and cautious evidence ought to have had greater weight with his hearers than their Report indicates. This Committee appears to have examined witnesses with the main purpose of getting evidence in favour of a foregone conclusion, namely, the need of enacting a law for the commitment and incarceration of habitual drunkards in special institutions called inebriate asylums, and indeed the Chairman himself asked one of the witnesses (Q. 462) whether he had "considered the proposition which was lying at the back of this Committee," namely, such an enactment.

Dr. Mitchell, however, plainly told the Committee

that "this proposed legislation does not deal with the causes of drunkenness; it deals with the other end of the evil;" and when asked (Q. 1211) by the Chairman whether he thought "that legislation of this kind would tend to diminish drunkenness throughout the country," Dr. Mitchell replied:—

"I do not think it would; such legislation as is here contemplated would not tend to diminish drunkenness, except perhaps by its indirect effect in making the young feel that it was disgraceful ever to be drunk, and a dangerous thing to be often drunk, as that might lead to compulsory work and loss of liberty. In such legislation as is contemplated at present, we are simply mitigating a mischief, the growth of which we have made no well-directed effort to check. I should like to see it a compulsory part of all education for which the State pays, that the young should be taught that it is their duty to understand the laws by which God governs the world, and to pay a reverential respect to them."

This is true wisdom, and would be right statescraft were it acted upon. Create a new sentiment among the people with regard to this most mischievous and degrading vice, and such a change of conduct will take place in the substratum of society as within a few years we have seen in its surface layer. If from public funds we were to create inebriate asylums for the drunken masses, "we should ruin the sober and well-doing, the class is so large." We should, moreover, teach the pernicious doctrine that drunkenness is an uncontrollable morbid impulse, to be cured by treatment in a kind of hospital, and therefore that it is not a degrading vice to be resisted in its first beginnings, or to be overcome by resolute effort in its progress.

With regard to upper-class drunkards, for whose benefit public time and money was mainly spent by the Parliamentary Committee, it needs some patience to consider calmly all the maudlin sentiment which is written about them. Of late years the upper class of English has become sober, and its growing opinion stamps drunkenness more and more as a disgrace; and that some small proportion of its members are left behind in the shameful indulgence of the old vice is certainly not a matter of national concern. But they will ruin themselves! No doubt, and why should they not? Their possessions will be better placed in sober hands, and their undeserved social position will be yielded to the advance of more worthy candidates. But they will kill themselves! And this also is more likely than lamentable, especially if they leave no offspring to inherit the curse of their qualities. It would be a national, nay a world-wide blessing if alcohol were really the active poison which it is so often represented to be, that men who indulge in it might die off quickly. The French have somewhat improved upon pure spirit in this direction by the invention of absinthe, which causes epilepsy, and the Americans, with their vile compounds of raw whisky taken into empty stomachs, are far ahead of ourselves. An American drunkard who sticks to his work has a much better prospect of finishing it within a reasonably short time than the Englishman, whose usual habit it is to drink less poisonous liquor with or after food.

And if habitual drunkenness is so inveterate a vice that the Scotch Commissioners, in all their vast experience, have never met with a satisfactory instance of reform, is it not better, on all hands, that it should run a short than a long course? The thorough-going drunkard soon puts an end to his worthless existence, and there the evil stops; but he who prolongs the agony remains for an indefinite number of years a disgrace to his people and a danger to society, and, worse than all, sows the foul seed of hereditary mischief.

Happily, drunkenness is a direct cause of sterility, to the extent of two-thirds of the children who would otherwise be born to sober parents, according to Lippich. But when drunkards do commit the *crime*, as Mill calls it, "of bestowing a life" which certainly has not "the ordinary chances of a desirable existence," the beneficent law of nature steps in to prevent the permanent degradation of the race. Morel of Rouen has shown, in his remarkable work on "Les Dégénérescences Humaines," how surely the race of the drunkard dies out in two or three generations after passing through the phases of nervous decay which we recognise as hereditary insanity and sterile idiocy. It is a new and awful example of the great conservative law of the Survival of the Fittest.

But the desire is uncontrollable, and therefore the indulgence is innocent.

This may be said of all desire which culminates in conduct; and in a certain sense, all desire is uncontrollable which is not controlled. Hence the sin is really in the mind, not in the action. But is the desire for drink uncontrollable? Wonderful stories have descended to us from Marc, Macnish, and other old authors as to drunkards preferring a plunge into Tartarus to abstention from one draught of brandy; and perhaps they are truly recorded. But to a drunkard Tartarus seems a long way off, and perhaps he thinks it will not be very hot there after all. Moreover it does not cost him much to say what he would dare to do. Has any one tried in vain upon a drunkard the certain infliction of an immediate pain, or the refusal of a pleasure greater than his drink? I can give a case in which the pleasure was not apparently very great. Many old Rugbeians will remember, as I do, J. S., the clever, amusing drunkard, who used to entertain their boyhood with music and legerdemain in the dining halls. After a heavy debauch he made a bet of one guinea with Mr. Sam Bucknill, the old school doctor, that he would not get drunk again for a twelvemonth, and he won it. He waited until midnight of the last day of his sobriety, and then steadily recommenced the process of drinking himself into his grave. He was never sober again.

There is, however, one practical, business-like point of view from which the agitation for a law of compulsory detention in inebriate asylums has arisen. We learn from Dr. Peddie's paper read before the last meeting of the British Medical Association that these institutions under the present law are *not remunerative*, because the inmates will not remain in them a sufficient length of time to make them so. This fact, that they do not pay, has prevented such a liberal expenditure as would make them attractive; for Dr. Peddie thinks that the bodily comforts and the amusements of the incarcerated drunkards ought to be very carefully provided for, "pleasing the palate with good culinary arrangements," "inducements for sport in the way of fishing and shooting," and the prison bars generally gilded as much as possible. If Parliament will only be so good as to pass a law which will sanction "compulsory commitments" and "prolonged detentions," then these enterprises would come to resemble the "many existing and thriving lunatic retreats."

This is outspoken at least, but whether Parliament

will come to the rescue "of inebriate institutions which have struggled under cramping difficulties" by adopting a new principle of criminal legislation seems more than doubtful. The knot seems scarcely worthy of such unravelment. It was unfortunate that Dr. Dalrymple, earnest and honest as he was, should have visited the United States to study the operation of the inebriate laws at the time he did. And still more so that his Committee should have been so much led by the evidence of his American witnesses. Even that cautious and enlightened statesman, Dr. Playfair, has admitted that his opinion, originally hostile to the proposed law, was changed by the evidence he heard, although I think at the present time he holds his judgment in suspense. When Mr. Dalrymple visited the States, the inebriate asylums there possessed to some extent the confidence of the public. And the Empire State and the Keystone State, and also the city of New York even, possessed such institutions of their own at Binghampton and Media and Ward's Island, supported partly out of public funds. Moreover, State laws had been enacted and were in operation, authorizing the commitment and compulsory detention of habitual drunkards in these asylums. Moreover, the Government of Upper Canada had built, or were building, an inebriate asylum at Hamilton, and had passed an Act similar to the New York Act to authorize commitment and detention therein. The inebriate asylum movement was in the early swing of apparent success, and the statements of results were readily accepted from the enthusiastic but interested men of whose worth and work they were a measure. I have carefully read the evidence which Mr. Dalrymple gave before his own

Committee, and I am bound to say that under the circumstances in which he saw the inebriate institutions of America, it seems to me eminently fair and impartial. It could not fail to have an immense effect upon the minds of the other members of his Committee. But, in addition, the Committee examined at great length Dr. Parrish, the superintendent of the Pennsylvania State Asylum at Media, and Dr. Dodge, the superintendent of the New York State Asylum at Binghampton, who supplemented Mr. Dalrymple's evidence with statements of facts of which they professed to be cognizant, and especially with the very weighty fact that they absolutely cured about 34 per cent. of their drunkard patients, though, indeed, in one place Dr. Parrish says, "There is no such thing as a permanent cure of anything."

Last year I visited the United States, and found that the condition and position of inebriate asylums there had undergone a great change since Mr. Dalrymple's visit in 1871. First I will mention changes of public notoriety. The State Inebriate Asylum at Media had been suppressed because it was said to be a failure. The New York Inebriate Asylum in Ward's Island had been ordered to be suppressed for the same reason. At Binghampton Dr. Dodge had left, and Dr. Congdon was carrying on his work with no better success. The Government at Albany had consented to give the institution a trial for one more year, after which time, unless it redeemed its character, the building was to be devoted to other purposes. In consequence of the failure in the States, the Government of Upper Canada had converted the building they had erected for an inebriate asylum into a lunatic asylum, and they had repealed the statute for the control of inebriates.

The New York statute has not been repealed, but it has been decided in the Supreme Court that it is unconstitutional, and it is therefore now never acted upon. The ground of this decision was that an intemperate American is a citizen not only of his own State, but of the United States, and that the State has no right to deprive a citizen of the United States of his liberty for conduct which is not criminal. In America the law is strained to permit drunkards who are not insane to be sent to asylums. In Maryland they are classified in a ward apart, and Dr. Conrad, the superintendent of the State Hospital there, said in a public discussion at which I was present—

"We have one hall [ward] devoted to inebriates or dipsomaniacs. The experience which I have had in the hospital has been confined to a class known as dipsomaniacs. Many have been coming to the hospital for several years, scarcely making an endeavour to withhold from drinking three days. *I do not know of a single case where a cure has been effected by confinement.*"

My own impressions of the inebriate asylums of America—and I visited six of them—are most unfavourable. I believe the treatment of habitual drunkards for the cure of their supposed disease to be unsound from top to bottom and everywhere. I make no exception; for the only institution in which I did find good, honest, earnest work being done was the inebriate *Reformatory* at Philadelphia, in the management of which the idea of curing a disease is steadfastly put on one side. All honour is due to the devoted men and women who labour in this place at the regeneration of their fallen fellow-citizens. But elsewhere I saw and heard nothing to show that any earnest effort was being made to change the habits of the inmates, or that the so-called institutions were anything more or better than boarding-houses within the walls of which the open consumption of strong drink was discountenanced; "capital places to pick up in after a debauch," as more than one inmate told me; "but good for nothing else." All the inmates whom I questioned admitted that without the slightest difficulty they could obtain what drink they liked during their daily walks, and with still greater freedom some of them ridiculed the supposed restraints of other institutions through which they had previously passed. It was not one of the labours of Hercules even to have a private store of drink within the asylum. The physician to a neighbouring hospital told me that on the occasion of some private theatricals at the inebriate asylum for the city of New York, he visited four of the inmates in their own rooms, and each one of them was able to offer him the choice of spirits out of his own cupboard. In fact, these American inebriate asylums seemed in some way to be part of the great whisky fraud.

At Binghampton especially, which I visited in company with Dr. John Gray of Utica, and Dr. Burr, one of the governors, the utter hollowness, or rather the total absence of any attempt at discipline and treatment was most obvious. One young man I did find shut up in a small room in the basement, and I was told that he had been there for ten days for repeated acts of drunkenness and for abusing the doctor. His friends refused to remove him, and the doctor hesitated to turn him out of doors penniless. But the other inmates, whom I found an educated, intelligent set of men, spoke very freely of the absence of all restraint. I was also told that the conversation and behaviour of the inmates, most of whom had led dissolute lives, was far more amusing than edifying. On the whole I came to the conclusion, that if Binghampton does cure 34 per cent. of habitual drunkards, the air of the place must be remarkably salubrious.

I heard also some account of those domestic mischiefs which Bentham anticipated in the passage I have quoted; of intemperate husbands who had been robbed of liberty that their wives might enjoy licence, and of other abuses which the evil passions of men and women are sure to suggest, with such a facile instrument as *lettres de cachet* for an imputed vice, for they referred to the time when committals were in force. On going through these institutions my mind was strongly moved with the inward question— How am I to know that these people are habitual drunkards or drunkards at all? In a lunatic asylum I can pick out any sane man who is wrongfully detained, but these men differ in nothing from myself except that they are said to be more prone to yield to a certain temptation. If compulsory detention be permitted, will it not often be used most wrongfully? for no one can distinguish the right from the wrong persons to whom it may be applied. But even if these places had been as successful as they have been the opposite, what sufficient precedent could they afford for a grave change in our law? They are occupied by a small number of inmates drawn from the well-to-do classes of society, of whom Mr. Dalrymple says, "The thing which struck me very forcibly was the *exceeding luxury* of the better class of patients but then they pay £10 a week."

Probably all the inebriate asylums of the United States do not contain 250 inmates, a very small datum to draw heavy conclusions from. We have larger material than that in the inebriate asylums of our own country, from which we might, if we took the trouble, obtain trustworthy statistics. I do not know what the statistics of permanent reform in them may be, although, judging from the freely expressed opinions of medical men, they are by no means encouraging. The superintendent of one of the best of them, however, let me a little behind the scene as to what became of his old patients. He told me that the number of patients who passed through his house during the year was considerable, that not a week passed without patients going out and new ones coming in, and that of those whose history he had been able to follow, a very large proportion died within two years; and this is exactly what a physician would expect.

Finally, I think we have no data which would justify us in appealing to the legislature for a new law, which would curtail in a most anomalous manner the liberty of the subject, on the plea of promoting the cure of habitual drunkenness. I do not think this fear of interference with the liberty of the subject "balderdash," as it was called at the British Medical Association meeting. Habitual drunkenness which can be distinguished as a form of mental disease can be dealt with under the lunacy laws. A London physician told the Parliamentary Committee that he had a great many such patients in his private asylum. "Perhaps there had been a little undue straining of the law to receive them," but there had been "faint scintillations of aberration," &c.,

which justified the medical man in certifying the insanity.

There need, however, be no straining of the lunacy laws when any real symptoms of insanity coexist with habitual drunkenness; though as a matter of convenience, regarding the detail of treatment, such cases might perhaps be classified apart, either in separate wards or in a separate asylum.

It is no part of the duty of the State to deal by penal enactments with intentions and dispositions, and therefore, in dealing with drunkenness, it can only regard the overt act. Mr. Hill, the late recorder of Birmingham, once proposed that it should be made legal for the police to secure and detain all the well-known habitual criminals of that town, forming an extremely small fractional part of its inhabitants, by which preventive measure he showed that crime with all its consequences could be almost abolished; but the mischief and danger of punishing a man for what he might do in the future was at once recognized.

The overt acts of the drunkard ought to be punished in such a way as to make them a real warning, and especially the act of public drunkenness, which is a kind of indecent exposure; also failure through drunkenness to maintain children, and indeed all drunken conduct which invades the rights of others; and there can be no just reason why the punishment for such acts should not be accumulative. It is unreasonable that magistrates should have to commit the same person from fifty to a hundred times for a constantly-repeated offence, and the remedy would appear to be a *penitentiary* for habitual drunkard offenders, in which they should be compelled to earn their maintenance, and from which they should be released *on trial*, and live for a time under the surveillance of the police.

Until some evidence has been procured that adult drunkards are capable of being reformed, all public money expended on *reformatories* ought to be devoted to the reformation of the young whose vices and crimes are to so great an extent the effect of parental intemperance.

Although the duty of the State does not extend to the punishment of private and self-regarding vice, it is bound to prevent public temptation to vice. It cannot prevent incontinence, but it suppresses brothels; it cannot stop gambling, but it closes betting-houses. So, therefore, it seems to be the bounden duty of the State to place the sale of strong drink under stringent regulation; to the effect that the trader in drink may not be the pander of drunkenness.

Moreover, the ruling powers of the State ought to enforce the equable administration of the law, so that it shall not be overstrained in Birmingham nor relaxed in Glasgow, and drunkenness made the special curse of a locality.

But, above all, the opinion and the influence of each right-thinking and right-feeling individual member of the aggregate which forms the State ought to be and must be brought to bear against this grievous evil. And when we consider the immense change in public opinion in this respect since the days of our youth; when we see the clergy of all creeds, from the Catholic cardinal to the common ranter, for the first time in our history earnestly denouncing the drunkard as a miserable sinner; when we see gentlemen regarding the vice which was fashionable with their fathers as the extreme mark of vulgarity; when the legislature has alreadly stamped drunkenness with the fitting sign of disgrace, by placing the policeman's hand upon the shoulder of any sot reeling along the highway; when the professed enemies of drink are enrolled in an army of three millions; and the foundations of a national education have at last been laid, we may feel assured of the steady increase of that national opinion, the only soil in which a national temperance can take firm root.

It is not drunkenness we wish to punish, but temperance we wish to promote; and to conclude, as I began, with Milton's words, "Were I the chooser, a dram of well-doing should be preferred before many times as much the forcible hindrance of evil-doing."

JOHN CHARLES BUCKNILL.

IV.

ON SOME RELATIONS BETWEEN INTEMPERANCE AND INSANITY.[1]

OUR worthy Secretary having called upon me, at a somewhat late hour, to furnish you with a subject of discussion this evening, I have chosen the very important subject of alcoholic insanity, upon which, as it seems to me, our knowledge remains in a very ill-sorted and unsatisfactory condition.

Permit me, first, to submit to you certain propositions as to pathological varieties of drink-insanity, which seem to me well worthy of your attentive consideration. I wish to avoid all reference to that dreadful slogan, the classification of insanity; but I must remind you that a collection of diseased organisms is not like a piece of slate rock, with cleavage only in one direction. You may very fairly divide them lengthwise, or crosswise, or cornerwise, and tie them up and deposit them in any variety of parcels or mental pigeon-holes which may please you or seem to instruct you.

Now, it seems to me that the commonly accepted forms of insanity from drink, fairly well marked as

[1] Read before the Medico-Psychological Association, January 31, 1877.

they doubtless are in typical instances, have yet no sufficiently pathological foundation upon which to base a just conception of the real relationship of drink and insanity; and that, moreover, they do not adapt themselves to the generalisations I am able to make from my own observation and experience. These forms, as we all know, are: 1. Delirium tremens; 2. Mania *è potu*; 3. Dementia *è potu*, or alcoholismus chronicus; 4. The moral insanity of drink, or oinomania.

No doubt a large proportion of the cases that we meet with may be classed under one or other of these four heads; but it is one of the questions upon which I most desire to compare your experience with my own, whether these forms are not too definite and restricted, and whether we do not meet with cases of insanity caused by drink which present to us every known variety of mental disease. My own not very narrow experience would affirm that it is so; and that, although a certain concourse of symptoms appertains to a large proportion of cases of drink madness, yet there is a wide margin of cases which are strangely varied and quite unimpressible within the forms I have named.

If this be so, some new attempt to devise a scheme of drink cases must be made, and I shall venture to propose one in correspondence with my own clinical experience.

First, I think we have a number of cases of which mania *è potu*, or acute mania from drink, is the type, in which the alcohol acts as an excitant of morbid cerebral function. In these cases, there is almost invariably a strong hereditary tendency, or a previous history, of brain-disease; and the alcohol lights a train of mischief already laid, and which might have been exploded by any other moral or physical cause of mental excitement : by a fit of passion, or by mental overstrain, or by a blow on the head, for instance. In these cases, the mental symptoms may be maniacal, or melancholic, or delusional ; in fact, they may be anything within the range of mental disturbance, be anything for which the brain was prepared ; but it is to be noted that they are not characterized by the marks of the second mode in which alcohol acts, to which I shall now advert. These are the cases in which the alcohol acts, not as a cerebral excitant, but as a neural poison, producing a characteristic group of symptoms, varied in intensity so as to justify subdivisions, but rarely any of them altogether absent in some part of the course of each case. The typical form of this mode of causation is delirium tremens ; but there are also many chronic cases of insanity in which muscular tremors, hallucinations (especially of hearing), local anæsthesias and palsies, and delusions of suspicion, persecution, and fear, mark their common origin in the toxic action of alcohol.

If these cases survive and are not cured, they pass into the third form by losing the more pronounced signs of nerve-poisoning, and acquiring those of cerebral atrophy and mental decay. But not seldom cases are met with in which the primary symptoms of cerebral disease from drink are those of degraded nutrition. The alcohol has not been taken in sufficiently large doses to act as a poison, but, continuously taken in smaller doses, it has set up its peculiar action in the red blood-corpuscles, diminishing their functional activity as oxygen carriers, and, if Binz be correct, diminishing the movements of the white corpuscles on which tissue-growth depends; and it has also acted directly on the cerebral capillaries, directly or through the vaso-motor centres, dilating them, thickening their coats, and deteriorating their function, the result being cerebral atrophy and mental decay. We may call these three modes Alcoholismus excitans, Alcoholismus toxicus, and Alcoholismus atrophicus.

I have yet to mention a fourth mode of drink-madness, which, if it do really exist as a form of mental disease, is a very distinct and remarkable one —I mean the moral insanity of intoxication, or oinomania. It is said that this form of insanity may be caused by blows upon the head, by hæmorrhages, by hereditary transmission, and other influences not alcoholic, so that to call it an insanity caused by drink would, in such cases, be incorrect. Other instances, where the craving for drink is the result of a vicious habit merely, cannot rightly be called cases of insanity until the alcohol has produced its effects upon the brain, either as a poison or as a denutrient, to the extent of developing the signs of mental derangement; after which they will be included either in the second or the third of my specified modes. Therefore, I exclude oinomania from drink-madness; and if it do really exist, I place it with its congeners, kleptomania and homicidal mania, in the class of moral insanities.

Now, the existence of kleptomania or of homicidal mania would never be admitted simply on the evidence that theft or murder had been committed. The vicious circle would be too apparent for any one to be permitted to argue or to plead, he is mad because he has stolen, and he has stolen because he is mad. In all cases of moral insanity, a group of symptoms must be estimated in their collective significance; and if such a group of symptoms as that, for instance, by which Anstie depicts oinomania, in Russell Reynolds's *System of Medicine*, be considered as necessary qualities or attributes of oinomania, I admit that such cases are met with and are rightly described; and I have only one further remark to make respecting them, namely, that they are veritable and mostly incurable cases of insanity, and that they may without difficulty be dealt with under the existing lunacy laws.

We may now pass to another but allied subject, namely, the Statistic of Drink-madness, and the right method of investigating it. As we all know, the old method has been to lump together all insane persons of whom it was reported that they had been drinking, as cases of insanity caused by intemperance. Even this rough method failed to justify the statements with which we are but too familiar, namely, that drink causes from 50 to 80 per cent. of all the mental disease in this country. But, although the enthusiasts of total abstinence may, without much blame, be careless arithmeticians, for the purposes of science a stricter method and a more accurate estimate of the drink-agency are essential; and a more discriminating investigation has already been instituted, at my request, by several of my friends, whose statistic I have already published in the July number of the *Journal of Mental Science*; and I have now an important addition to make, for which I am indebted to my friend Dr. Major, respecting the drink-statistic of the 511 cases of insanity admitted into the Asylum for the West Riding during last year. Of these, 11·35 per cent. were Class A cases of insanity, resulting from the direct action of alcohol; 1·56 per cent. were Class B cases, complicated by hereditary tendency to insanity; 2·93 per cent. were Class C cases, in which alcoholic excess had been combined with other adverse physical conditions; and 1·95 per cent. were Class D cases, in which alcoholic excess had been combined with mental causes—making a total of 17·79 per cent. of alcoholic cases on the admissions. The total percentage on the male admissions was 31·20, and that on the female admissions only 4·98.[1]

In forming a sound statistic on this subject, the first requisite is to exclude all cases in which drunkenness is not a cause but a symptom of insanity, which all experienced alienists know to be frequently the fact; for it would be as irrational not to make this distinction as it would be to lump together all insane persons who have suffered losses in business, not distinguishing those in whom the loss had been the

---

[1] Dr. Rhys Williams has since kindly provided me with the following important statistic of drunkard cases admitted into Bethlehem Hospital during the two last years, arranged in the same four classes.

	Males.	Females.
Number admitted during 1875	94	138
Number admitted during 1876	114	129

	Number.		Percentage.	
	Males.	Females.	Males.	Females.
1875.—Class A.	2	2	2·19	1·44
,, Class B.	—	4	—	2·89
,, Class C.	1	2	0·10	1·44
,, Class D.	4	1	4·36	0·72
1876.—Class A.	2	2	1·75	1·55
,, Class B.	2	2	1·75	1·55
,, Class C.	—	—	—	—
,, Class D.	2	2	1·75	1·55
	13	15		

	Males.	Females.
Total number admitted in 1875 and 1876	208	267
Percentage of cases attributed to intemperance	6·25	5·61

INTEMPERANCE AND INSANITY. 87

cause of the insanity from those in whom insanity had been the cause of the loss.

Moreover, it is essential to a right understanding of the drink-etiology of insanity that the cases which have been directly caused by strong drink alone should be separated from those in which alcohol has only acted the part of an ally with other enemies of mental health; for it is very certain that, in many cases in which intemperance has aided in this warfare, its share of influence has been by no means the greatest; hereditary predisposition, mental overstrain, worry, and an array of combining causes having had by far the greatest power in bringing about the common result. The truth and importance of this view seem so obvious that I may well spare you further comment upon it, with the expression of an earnest hope that, now it has been accepted and carried into practice by our eminent associates, Drs. Clouston, Major, Duckworth Williams, and your President, this more discriminating statistic of drink-madness will be generally attempted in all our institutions for the insane. In carrying it out, I think some difference of opinion and practice may arise in the discrimination of the direct from the indirect cases; and I am myself inclined to believe that, the more thoroughly the histories of our drink cases are investigated and pondered upon, the larger will be the proportion of them which will be shunted from the direct to the indirect line of causation.

And now I have to offer to you a consideration, which will, at first sight perhaps, have a paradoxical ring, as coming from me after all that I have said and written respecting the potent influence of alcohol in the causation of insanity; but it is just in a careful analysis where the remainder is least forgotten. I am warned, however, by old experience, of the way in which one's words are apt to be perverted from their true meaning, to protect myself by avowing once again that no one can detest drunkenness more than I do; that I think it the greatest remaining curse of this age and country; and that I believe intemperance in drink to be by far the most potent of all removable causes of mental disease.

After this, I may venture to indicate what I think to be another equally real aspect of drink in relation to insanity, namely, the causal relation between the occasional use of alcohol and the prevention or postponement of mental disease.

With men of such wide experience as my present audience, a few considerations will probably suffice to gain me many suffrages in favour of this novel and, I fear, startling proposition; but let us bear in mind many of the commoner moral and physical causes of insanity, the prevailing bodily conditions of the incipient disease, and the necessities of the treatment, and we must, I think, see and admit that this stimulant-narcotic, in such general use, must have a vast and varying influence upon the organisms of men, which is not likely to be invariably pernicious, and which may well be sometimes beneficial and conservative of the mental health.

Consider the great part which grief and anxiety, worry and overstrain, play in the production of insanity, the depressing effects of poverty and the failing struggle for existence, of misery in all its forms, and then consider to how great an extent the use of alcohol oftentimes tends to make the burthen of life bearable, if not by stimulating the powers, at least by deadening the sensibilities of men; and I think you will agree with me that, by the occasional help of strong drink, a man may sometimes be able to weather that point of wretchedness upon which his sanity would otherwise have been wrecked. The observation of life forbids us to doubt that "wine, which cheereth God and man," according to Holy Writ, doth sometimes blunt the keen edge of misery, so that the wretch is not "cut to the brain," like King Lear. Alcohol, in its physiological action, is *atriptic*, retarding the disintegration of the tissues, especially of the nerve tissue; and, when the brain is wearing itself into madness, alcohol, at the right time and in the right dose, does without doubt sometimes check the ebb-tide of reason. Perhaps, a few timely doses of opium might have the same or a better result, if the people of this country were in the habit of resorting to opium to dull their misery and assuage their pain; and in China opium, although the source of infinite mischief, is also, no doubt, a precious boon to the miserable who may use it aright, either by happy chance or wise direction.

Alcohol, moreover, is not only a narcotic which may "knit up the ravelled sleave of care;" it is also, according to Anstie, Lauder Brunton, and all good authorities, a food, and as such it plays an important part in the therapeutics of insanity. I have myself no doubt that a moderate use of fermented drink is useful in the treatment of mental disease, not only that a cure, when possible, may be attained *cito, certo, et jucundè*, but that, in incurable cases, the bodily health may be improved and the mental misery alleviated.

I do not wish you to infer that, if called upon to

adopt prophylactic treatment in a case where insanity was threatened, I should be likely to prescribe alcohol in any form, unless it were specially indicated. I have generally succeeded in finding some better method of escaping the danger. But our uninstructed countrymen, whose custom it is to drown their phrenalgias in the flowing bowl, do, according to my observation, sometimes succeed in the dangerous enterprise. The dreadful mischief of which strong drink is the source, in the causation of insanity, affords no good reason why we should refuse to observe that, under exceptional circumstances, it has no slight influence in the prevention of the same disease. *Ubi virus, ibi virtus.* Statistics, indeed, supply us with no ready-reckoner of this last-named result; for, in the words of one who enjoyed much and suffered much,

> "What's done ye aiblins may compute,
> But never what's resisted."

Burns, had he been sober, might have made a fortune by industry, lost it by speculation, and ended his days in the Crichton Asylum—leaving the world all the poorer for the want of some of its sweetest and tenderest poetry.

To conclude: as alcohol, by causing partial paralysis of the nervous mechanism, will sometimes obtund the shock of physical injury, which would otherwise be fatal, so, in like manner, it will deaden the blow of mental pain, which would otherwise destroy the reason But, remember, that it is but a precarious refuge on urgent occasion from a greater danger than its use itself involves. The mental physician loves not an alcoholised patient one whit better than the surgeon does; for in either case, the repetition of the remedy may speedily brew as much mischief as the original injury could effect, and the man who resorts to the bottle to drown his habitual cares is on the downward slope of a road which surely leads to perdition.

Let me not, therefore, be misrepresented as recommending strong drink as a remedy for grief and care when they threaten sanity, because I say that, owing to the customs of living in this country, it not unfrequently is such a remedy, although a most unsafe one. But we physicians are not fanatics, and can have no more antipathy to alcohol than we have to arsenic, when used aright for a beneficial purpose; and we ought not to be debarred from recognising such possible occasions of its use, because more commonly it enters, as an evil spirit, into a herd of human swine seeking their own destruction.[1]

[1] The Report just issued by the Registrar-General contains a most interesting letter from Dr. W. Farr, in which he comments on the action of alcohol as a preventive of disease. He says:—"Alcohol appears to arrest the action of zymotic diseases; as it prevents weak wines from fermenting. Like camphor alcohol preserves animal matter; this is not now disputed. But may it not do more? May it not prevent the invasion of some kinds of zymotic diseases? I invite the attention of those who have portrayed the bad effects of alcohol, to consider whether it does not prevent the action of various infections in the temperate. The neglect of this side of the question throws a doubt upon many of their inferences. The deaths attributed to zymotic disease in 1876 were 96,660, to alcoholism, 1120; now it is evident that any effect depressing the prevalence of zymotic diseases that kill their tens of thousands will save the lives of thousands. . . . Dr. Parkes's careful experiments were made on a soldier not in company, and the effects on his mind were not noted; yet that is more striking and important than the effect upon temperature, and on the secretions. The effect on the brain stands before that on the heart."

V.

To the EDITOR *of the* "TIMES."

SIR,—

IN the letter on habitual drunkards which appeared in the *Times* of Friday, Oct. 20, Dr. Alfred Carpenter complains that "the time of the police is taken up by a necessary attendance upon these poor unfortunates, while criminals who should be carefully watched are enabled to perpetrate their burglaries undetected."

As, however, burglars are a smaller and more dangerous class even than habitual drunkards, would it not be the more simple method for the police to apprehend and incarcerate all habitual burglars, so that they might have time to watch the drunkards?

Some years ago the Recorder of Birmingham strongly and persistently advocated a law for the apprehension and detention of the whole class of habitual criminals in that town, thus saving taxes and trouble to an extent which seemed almost to justify that bold innovation. The only possible objection to it seemed to be that it would be contrary to the spirit of our laws to punish any man in anticipation of any offence which he is only expected to commit; and this objection applies with quite as much force to habitual drunkards as to habitual criminals, for no man can predict with certainty that either the criminal or the drunkard will persist in his previous conduct.

Dr. Carpenter, indeed, denies that his desire to lock up for a long time men who are addicted to drink is conceived with a punitive purpose, or in a vindictive spirit for the punishment of wrong-doing; but the purpose of the man who turns the key will never sweeten imprisonment, and it is to be hoped that even for offences actually committed our magistrates do not employ the services of the gaoler in a vindictive spirit. Dr. Carpenter asserts that the power of imprisoning persons addicted to drink is not a new one in principle, seeing that it already exists as regards lunatics and children who have been convicted of crime.

But the cases are not parallel. Lunatics who are deprived of liberty for convenience of care and treatment and for the safety of society, can only be legally detained so long as they are lunatics. When they recover the possession of their faculties they cannot be longer detained because they are liable to become insane again, and they must be discharged. The State provides that lunatic asylums shall be visited by officials who are capable of saying, with a large amount of certitude, whether each inmate is or is not actually insane, and therefore properly detained; but such discrimination is and must be utterly impossible in an inebriate asylum, in which the inmates are all sober and to all appearance exactly like other people.

In juvenile reformatories the State places itself *in loco parentis*, and is justified in its abnormal action by the complete reform of a very large proportion of neglected children—an experience which no candid and well-informed person can attribute to the treatment of adult drunkards.

Evidence was given before Mr. Dalrymple's Committee that the compulsory detention of drunkards for any period short of three years was quite useless; and Dr. Carpenter would even seem to desire an extension of that heavy sentence, seeing that he advocates the detention of the drinker-suspect "until the time arrives at which all the tissues of his body shall have been changed, and a new tissue laid down in its place." Unfortunately, new tissue is laid down in the mould and on the lines of the old tissue, so that there is no reason to expect that a man's character will be changed by the renovation of all the tissues of his body. The elimination from the tissues of the drink poison only requires quite a brief abstinence, which has been experimentally proved by Magnan to be not less than three and not more than seven days. This elimination and its attendant sobriety would correspond with the cure of a lunatic, and no medical reason can be given why the treatment of mere drunkenness, independently of its remote effects, should be extended beyond it.

I very much question the accuracy of Dr. Carpenter's statement that the almost unanimous opinion of the medical profession is in favour of a new law for the imprisonment of persons addicted to drink. The real fact seems to be that many medical men, but not all, have hitherto omitted to consider the question seriously, but, like Mr. Dalrymple's Committee and the public generally, have been satisfied to accept, without investigation, the statements of American enthusiasts, which are now utterly discredited in their own country. Thackeray, I think it was, who said that he was always puzzled to know what became of English scarecrows, until, in the west of Ireland, he found the peasantry dressed in them ; and so we find the rags and tatters of opinions which have become scarecrows in the United States afford drapery and ornament to the promoters of social science among ourselves. The women's rights question and spiritualism are cases in point, but the newest importation is this habitual drunkards movement.

I dare not venture to trespass upon your space by describing the disreputable condition of the American inebriate asylums which I myself witnessed last year, but I may be permitted to state a few pregnant facts indicating the present condition of American opinion regarding them.

The Pennsylvania State Inebriate Asylum, of which Dr. Parrish, *facile princeps* of Mr. Dalrymple's evidence, was superintendent, had, I found, been abolished as a failure.

The New York City Inebriate Asylum had been condemned and was to be abolished in a month.

The New York State Inebriate Asylum at Binghampton, of which the superintendent, Dr. Dodge, was the other bell-rope of American evidence before Mr. Dalrymple's Committee, was under trial for one year, with intimation from the Legislature that at the end of that time it also would be abolished, unless good cause to the contrary could, in the meanwhile, be shown.

The building for an inebriate asylum which had been erected by the Government of Ontario, had been appropriated to the purposes of a lunatic asylum in consequence of the complete failure which the Government considered that inebriate asylums had proved to be in the States.

Moreover, the New York State law empowering certain authorities to commit persons addicted to drink to the State Inebriate Asylum had not, indeed, been repealed, but it had been rendered quite inoperative and effete by a decision of the Supreme Court that it was unconstitutional, inasmuch as it was held that no State was competent to pass a law whereby a man being a citizen not only of the particular State, but also of the United States, could be deprived of his liberty, except for the commission of a crime, under which designation the tendency to make himself drunk could not be brought.

This judicial decision does not appear to have been come to on theoretical grounds alone, but in consequence of great abuses to which the law had led. I was myself informed of cases in which drinking husbands had been imprisoned under it by immoral wives, and troublesome sons by negligent parents.

The advocates for a new law for locking up drunkards cannot be permitted to throw overboard their Jonah of American experience when the wind of inquiry arises; for the conclusions of Mr. Dalrymple's Committee, so constantly referred to, were mainly founded upon American evidence; and if that evidence be discredited and disallowed, we have no base of knowledge of our own upon which to build such a law.

Would it not be much better to try the effect of an earnest and impartial application of our own law as it at present exists?

I am, Sir, your obedient servant,

JOHN C. BUCKNILL.

39, WIMPOLE STREET, *November* 3, 1876.

VI.

To the EDITOR of the "TIMES."

Sir,—

On the second reading of the Habitual Drunkards Bill the introducer referred to articles which I have written on this subject in terms which compel me to request the favour of putting myself right with the public by a brief explanation in your powerful columns.

In the first place, allow me to say that Dr. Cameron has not quoted my words correctly, and that the alterations he has made in them, no doubt inadvertently, have the effect of exaggerating my views.

Secondly, I have not, as Dr. Cameron asserts, written any articles with "strong prejudice," but with a painstaking investigation of facts, upon which I have arrived at opinions probably nearer to the Bill as it will pass than those of Dr. Cameron himself.

My statements with regard to the American asylums have not "called forth the most conclusive replies," if by replies Dr. Cameron means denials.

The only attempt at denial which I have met with was to my statement that the inebriates on Ward's

Island got easy access to whisky. It was said that this could not be, because there was no spirit store on the island; but my informant, Dr. Macdonald, who spoke from personal knowledge, explained to me that the boatmen well understood and obeyed the signals of the drunkard captives.

Dr. Cameron asserts that " an institution which he (Dr. Bucknill) reported closed through failure, was closed simply because the State aid that had been expected was not forthcoming." But if my information was correct, this State aid was refused because this institution was considered to have failed ; and, if so, I fail to see the distinction which Dr. Cameron attempts to draw.

Dr. Cameron's last accusation is, "that Dr. Bucknill's reports were founded upon superficial and limited information, and that he had not visited asylums the success of which had surpassed the wildest dreams of enthusiastic supporters." In reply to this, I have to state that I avoided one asylum, and one only, and I did so because I was forewarned that this was the very place where I should encounter these wildest dreams of enthusiasm, and Dr. Cameron's comment indicates that my information was correct. It was an asylum containing eight or ten patients only, kept by an estimable but enthusiastic man, whose friends, who were also my friends, did not hesitate to tell me that they had been grieved by his public statements. I am sorry to have annoyed him by passing him by, but if he should desire it I will give him the very words of an authority in America who is reverenced for practical philanthropy, which decided me to do so. With this single exception, . visited at no little cost of time and labour all the inebriate asylums which I could hear of in the Eastern States.

I will not prolong this personal explanation by any remarks upon what remains of Dr. Cameron's Bill, except by observing that I think it exceedingly unfortunate that it is so distinctively a class measure, and that as a justice of the peace I anticipate with some repugnance the duty of carrying out its provisions for treating the rich drunkard as if his conduct were the uncontrol'able result of disease, while upon the poor and ignorant wretch I must still impose the penalty of vicious excess.

I am, Sir, your very obedient servant,
JOHN CHARLES BUCKNILL, F.R.S.

HILMORTON HALL, RUGBY, *July* 6, 1878.

VII.

To the EDITOR *of the* "TIMES."

SIR,—

I trust that Dr. Cameron may see that my words are but the expression of the evolution theory by the elimination of the unfittest, a very different matter to the opinion that "the best thing that could happen was that the drunkard should ruin himself." I have still to complain of Dr. Cameron's inaccuracy. So far from having "denounced it (his Bill) when presiding over the psychological section of the British Medical Association last year, notwithstanding which denunciation the section proceeded to pass resolutions in favour of its provisions," I did not one jot more than impartially keep the discussion within bounds of order. After the votes were taken and the fight was done, I was jestingly taunted with having expressed no opinion and challenged to do so, and then it was that I let fly the sentence which has found a joint in Dr. Cameron's equanimity. The fate of Dr. Cameron's first Bill has proved the truth of my opinion, for had that Bill been wise, practicable, clever, and consistent, it would scarcely have been whittled down to the stump which remains.

So far from having "said not one word" about inebriate homes in America, in my first article I described and praised with all my heart the Franklin Home for the Reform of Inebriates at Philadelphia, to which I had paid repeated visits, and the imitation of which I strongly recommended in Glasgow and other centres of spirit-drinking at home. But this institution does not please the coterie of inebriate doctors, because it does not conform to their dogma that drunkenness is a disease to be cured and not a vice to be reformed. I have never asserted that 60 per cent., not even 34 per cent., which was then the claim, of the cases at Binghampton were not cured, although I heard the statement widely discredited by American physicians. What I did write was that I visited Binghampton in company with Dr. John Gray, the well-known superintendent of the lunatic asylum for the State of New York, together with two of the Governors of Binghampton, and that in the presence of these gentlemen, Dr. Congdon, the superintendent of Binghampton, fully admitted to me that he employed neither medical nor moral treatment of his patients, a disclosure which caused the Governors to induce Dr. Congdon shortly afterwards to resign, and which led me to the inference that the alleged cures, if true, were spontaneous. These facts were published by me two years ago, and they have never been denied.

If legislation is to be founded upon the assumed curability of drunkenness as a disease, would it not be prudent to obtain good evidence of the fact of home growth and some explanation of that evidence to the contrary which Dr. Cameron has ignored? In Scotland it has been the practice to detain habitual drunkards in lunatic asylums, thus bringing them under the observation of the Commissioners in Lunacy, who, however, state in their report for 1872 that "it would not be easy to point out one single case of permanent and satisfactory reform."

But surely if the medical treatment of drunkards be successful, Dr. Cameron would be wise to provide for it in institutions with salaried officials; for I know not on what principle the Legislature can sanction a new class of establishments in which any kind of imprisonment of the Queen's subjects can be carried on to the advantage and profit of private adventurers.

I am, Sir, your obedient servant,

JOHN CHARLES·BUCKNILL.

ATHENÆUM CLUB, *July* 12, 1878.

www.ingramcontent.com/pod-product-compliance
Lightning Source LLC
Chambersburg PA
CBHW022134160426
43197CB00009B/1275